Praise for *The Potentialist*

"Ben Lytle has written an important book. He reminds us that the only constant in life is change and that our ability to adapt and innovate will define our place in the world, both as individuals and as a nation. Based upon lessons learned as a successful businessman and community leader, Ben offers both practical and high-level insights worth the read for ordinary citizens, prospective entrepreneurs and policy makers alike."

EVAN BAYH
Attorney, advisor, board member, and former two-term governor and two-term US senator for Indiana

"There are many excellent books on the changes coming in the next few decades, but Ben Lytle is the first author to tackle in detail how those changes will affect our lives and careers and how to act now to adapt and thrive. Ben's perspective as a proven entrepreneur and leader at the highest levels of business provides the perfect roadmap for growth and success in the new era of business and life."

JOSH LINKNER
Five-time tech entrepreneur and *New York Times* bestselling author

"The world is changing quickly. Technology accelerates the divide between the people who create hardware and software and those who do not. Divides force people apart and politicians have been ineffective at bringing us together. Ben Lytle saw how technological advances could facilitate management of healthcare information and the reinvention of health insurance. Throughout his career, he has transformed organizations by mastery of change and informatics. His employees, customers, and shareholders—all ultimately 'patients'— have benefited. In *The Potentialist: Your Future in the New Reality of the Next Thirty Years*, he shares what will happen next and why. It's scary and exciting. He provides us with strategies to stop, think and flourish in a future world that prior generations could not even imagine."

WOODROW A. MYERS, JR., MD, MBA

Harvard- and Stanford-trained physician; former health commissioner for Indiana and New York City; former director of healthcare management of Ford Motor Company; former EVP and chief medical officer of WellPoint, Inc.; managing director of Myers Ventures, LLC

"Ben is someone who, simply put, never stops thinking about where we are and where we're going. He has made it his life's work to synthesize his knowledge and experiences into keen insights about the future. In *The Potentialist: Your Future in the New Reality of the Next Thirty Years*, Ben boldly lays out what the future will mean for the world of work and the expression of human potential."

JIM PARKER

Former Senior Advisor to the Secretary of the Department of Health & Human Services, health plan CEO, investor, and board member

"Ben's insights and experiences provide great wisdom and practical advice for navigating the coming years both from a professional and personal standpoint."

ROB ROSENBERG
Global Head of Human Resources
DHL Supply Chain

"Ben Lytle is a visionary who leads with intelligence, confidence, and humility. Who better to help write the operating manual for a thirty-year period of change for which there is no historical precedent? This book has startling revelations and depth, yet it oozes with hope—not fear. Ben posits that the choices we make now are not only smart but necessary; a chance to embrace the unparalleled opportunity in a brave new world of advanced technology without sacrificing our humanity along with way."

ANNE RYDER
Senior lecturer of journalism, Indiana University; Emmy Award-winning anchor/reporter

www.amplifypublishing.com

The Potentialist I: Your Future in the New Reality of the Next Thirty Years

For more information, please contact:
Amplify Publishing, an imprint of Amplify Publishing Group
620 Herndon Parkway, Suite 320
Herndon, VA 20170
info@amplifypublishing.com

Library of Congress Control Number: 2021923573

CPSIA Code: PRFRE0322A

ISBN-13: 978-1-63755-136-3

Printed in Canada

This book is dedicated to my eight grandchildren
and their peers worldwide who will lift humanity
to new heights and to those older persons who
encourage and inspire them to reach their potential.

The Potentialist

Your Future in the New Reality of the Next Thirty Years

Ben Lytle

amplify

CONTENTS

Preface

AT AGE TWENTY, A LIFE-ALTERING event started me on
a path different than most people follow, and certainly different
from what I had considered to that point. That event and its effects
led me to believe that the human race's best, and my personal best,
was achievable and more easily than commonly believed.

I was a Texas ranch kid raised by loving parents, siblings, and
a big extended family who encouraged me with statements like,
"You're a smart kid and you work hard, so you can do anything."
An inspiring and locally revered high school English teacher named
Opal Baker expanded my worldview beyond Texas with challenging
questions about life and an introduction to classic literature. She
asked me questions such as:

- "Do you see your future limited in any way, and, if so, what way?"
- "Do you believe that it is possible to make your dreams come true?"
- "You have a fine mind, a lot of energy, and love books and words. What will you do with that?"

She was the first person to explain the term *archetype* as patterns of human behavior that appear repeatedly in life and literature. She introduced me to Tolstoy, beginning with *War and Peace*. She followed up with a reading list that I was still working through a decade or more later. I sometimes return to that list today for a re-read. Like most kids in my small town, I worked multiple jobs from the time I was a preteen. My jobs included caring for the cattle on our ranch, repairing (*fixin'* in Texan) fences, delivering newspapers, working as a cotton picker, hay hauler (one of the nastiest, hardest jobs ever—if there is a hell, I'm convinced people haul hay there), pumping gasoline, flipping burgers, and lifeguarding each summer at the municipal pool. My best and longest-lasting job was working for a local wholesale-retail florist as a delivery truck driver and occasional long-haul driver on weekends. The owners took an interest in me. They encouraged my plans to work my way through college. They introduced me to white-collar work by allowing me to assist in the office. I took to it quickly. The owners implemented process improvements that I recommended even though I was only sixteen years old. During that same period, my mother encouraged me to consider a career in technology, which in those days was called "data processing."

As was common in my day, I married right out of high school and started a family. My new brother-in-law was a successful data processing executive. He made me aware of a job opening in data processing with the federal government if I could pass the entrance exam. I passed and got my first full-time job in this new, promising field. I thrived in my first three technology jobs and received a series of rapid management promotions despite my youth. The career paid well and enabled me to put myself and my new wife through college while working full time, being a father of two sons, and gaining invaluable technology and business experience. But despite all the family support, a better-than-average launch into adulthood, and a potentially lucrative technology career, I was clueless about what life meant or what to do with mine.

A defining moment arrived for me on a hot, humid August day on the campus of East Texas State University in Commerce, Texas, (now Texas A&M-Commerce), about an hour northeast of Dallas. A charismatic young instructor in a freshman elective class, intriguingly titled "Introduction to Personality," admonished those of us in his class to become captains of our fate using our personal potential as the North Star to guide us. He insisted that we need not live like the people before us or around us; we alone had the right and the responsibility to define our destiny. He explained that we should seek life success, not simply career success. Even though my relatives and the owners of the florist business had encouraged me to think outside of the box about my career and potential, it was liberating to gain the instructor's expansive perspective. He explained that much of what the ancient Greeks called *virtue*, what psychologists call *self-actualization*, and many religions encourage as a righteous life is rooted in or consistent with the pursuit of individual potential. He pointed out that over the centuries, millions of people had undoubtedly chosen pursuit of their potential as their life's work.

As a result of that young instructor's influence, I changed my degree from a major in business to psychology. (There were no technology degrees available at my college in those days.) I believed that developing my potential was more important to life success than anything I would learn in college about business or technology. As I learned more about the pursuit of potential in college and in the years after, it became a central theme of my life. I read extensively on the subject and invested in advanced classes and other self-study about what individual potential is, how to discover it, and what practices could be easily learned.

The instructor's succinct framework proved accurate over the years despite widely varying or nuanced interpretations and practices that I discovered in my research and studies. I learned that mentors, teachers, and guides for living to our potential are all around us if we look for them, ask for their help, and listen to their advice. Some of my teachers lived centuries ago but guided me through their writings. Others were contemporaries who taught me about business, family, career, and life. I will forever be indebted to novelists such as Tolstoy, Dostoevsky, and Hugo; philosophers such as St. Thomas Aquinas, Aristotle, and Emerson; and most importantly the influences of my parents, brother, mentors, adult children, and closest friends. Some of my most important mentors profoundly influenced me, including my first mentor-boss Larry Sweet who taught me how to lead people decades older than me when I was twenty-two years old; my boss Gene Hinkle who a decade later taught me how to be an executive in my early thirties; Lloyd J. Banks, who taught me to be a CEO while I was his Chief Operating Officer; and Dr. Sharad Desai, who, over the course of a decade, introduced me to the finer points of being a human being. There were dozens of other co-creators too numerous to list. I met them in board rooms, bars, civic projects, in their legislative or executive offices, socially, and in my travels around the world. All my co-creators enabled the incredibly rewarding life I have been able to live. The second book of The Potentialist series shares much of what I learned about becoming a Potentialist. I hope that my experiences can be useful to you in your personal growth and development. Living to our potential has never been more important than it will be in the turbulent times ahead.

We can learn from almost everyone around us if we pay attention, listen closely, and observe actions as well as their words. People who choose to live differently and to make a difference fascinate me. In this book, I refer to them as "exceptional." I have been fortunate to meet and learn from many exceptional people and to record their stories and perspectives. Many were successful in the eyes of the world, but many others were more quietly exceptional and known only to those who were fortunate enough to be in their orbit. All were exceptional because they seemed to naturally or purposefully utilize their innate potential to achieve for themselves while making the world and the people around them better.

In my early twenties, I became fascinated by the writings of futurists. Futurists probe the direction and potential of the human *species*. To me, the work of futurists is inseparable from my study of *individual* human potential, because each informs and enables the other. I have always been fascinated by history and anthropology because our evolution and the arc of history inform us about our collective past and what our collective future may hold. These subjects became lifetime avocations that I applied to my career and life. As a prolific note-taker and reader, I documented much of what I learned, initially in three-ring binders, and later in digital files.

A few years ago, I began to appreciate the magnitude of changes coming in the next three decades. I was concerned then, and remain concerned, that few people are aware of the degree of change coming to lives and careers, or how quickly they will need to adapt. I was most concerned for my three adult children, their spouses, and my eight grandchildren. However, by extension, I became concerned for everyone in their mid-fifties and younger.

As described in the Prologue and Part 1 (chapters 1 through 4), the pace of change and innovation will accelerate, exponentially challenging individuals and institutions of society to adapt as never before. The world and daily lives will feel more volatile because resistance and adversity accompanies change and innovation. Historical limits to human existence and potential will be reduced or disappear—such as the minds and bodies we were born with, our ability to communicate in any language, the distances that separate us, and the fear and consequences of overpopulation. They will be replaced by new challenges including invasions of privacy, security, and individuality, and adapting economic systems to declining populations. I began summarizing what I had learned about the near future, individual and collective human potential, and what can be learned from exceptional people. I attempted to reduce decades of study and observation into simplified concepts and practices that could be explained in everyday language. My initial intention was to create a white paper for my family. However, the white paper became a book, which turned into three books, the first of which you now hold in your hands.

The Potentialist: Your Future in the New Reality of the Next Thirty Years and the other two Potentialist books are the result of that condensation process, plus my experiences and point of view. I am a pragmatist, so all three books of the Potentialist series are ripe with suggestions for navigating the turbulent but opportunity rich times ahead. They are intended to stimulate your own unique search and decision-making. If you enjoy reading *Your Future in the New Reality of the Next Thirty Years* and would like to learn more about how to reach your potential, continue with books two and three in the Potentialist series and visit www.potentialistfuture.com.

I hope that you find this book and the rest of the series to be a useful starting point on your path to potential, success, expanded wisdom, and contributing to what will be a better world.

Through the Looking-Glass

> "Ray Kurzweil did the math and found that we're going to experience twenty thousand years of technological change over the next one hundred years." —Peter H. Diamandis, *The Future Is Faster Than You Think*

Looking back over the past thirty years, we've seen the dawn of the World Wide Web (later called the Internet), email and texting, smartphones, GPS, online video streaming, 3-D printing, and more. Life and its advances are always accompanied by resistance and adversity, so we also experienced a new era of terrorism with 9/11, the 2008 financial crisis, increasing political polarization, riots, and of course a crippling worldwide pandemic. In the midst of these volatile times, the institutions that have guided us for centuries such as employers; federal and local governments; and educational, religious, and social organizations have become less stable and reliable, as they too struggle to adapt to the rapid pace of change. Over the next few decades, some institutions and employers will become less relevant or fall by the wayside altogether. With fewer safety nets and less reliable institutional guidance, you will need to navigate this new world more wisely, independently and resiliently, relying less on their assistance. Simply stated, you will not be able to live, have a career, or retire in the same way as your parents and grandparents. You are entering a New Reality of human existence.

The period beginning with the COVID-19 pandemic and ending in 2050 will be a time like none other in history. The World Economic Forum refers to the coming decades as the Fourth Industrial Revolution. In my opinion, that is an understatement. The changes brought about by this revolution will be faster, more volatile, and farther-reaching. They will affect every person on earth. You will be required to make decisions that you never thought you would face.

I won't describe here the extensive list of innovations that will reshape the way we live over the next thirty years. There are dozens of books and periodicals—and thousands of news reports—that do that job well; many are listed in the extensive bibliography at the end of this book. Instead, I'll condense the multitude of changes you are going to face into three perspectives that vividly describe how you will be personally affected.

1. You will not be the same human being you are today. As a result of seamless integration of automation and medical science innovation, you will become something of a cyborg—part human and part machine, more powerful

and knowledgeable, living longer and healthier than humans have ever lived.

2. You will do more rewarding work over a longer life and in many more jobs and careers. You may choose never to retire as we know it today, or you may be forced to work longer to finance your extra years of life.

3. Most of your life will exist inside a digital world; the physical world will become secondary. In the digital world, you will work, shop, run your household, socialize, relax, and conduct commerce without the historical limitations of distance, language barriers, and birth origin (the ovarian lottery).

This is the future as forecast by a global community of scientists, technologists, entrepreneurs, and futurists. Some of the best minds and most progressive companies in the world are already making it a reality. None of this is science fiction—although at times it may appear that way—it is science *fact*. None of it is in the far-distant future. In fact, most of these developments will roll out in the next thirty years, and many in the next five to ten years. They will be upon you before you know it.

The big question is: *How can you prepare for this future?* What strategies can you employ today to ensure your success? I wrote *Your Future in the New Reality of the Next Thirty Years* as a "how to" guide to prepare you for an epic rollercoaster ride of unprecedented change, challenges, opportunities, and decisions. Let's begin that preparation by taking you through the looking glass ten, twenty, and thirty years into the future by examining the lives of everyday people. They could be your neighbors. One of them could be you.

∽

Alisha is fifty-six years old; her husband, Alejandro, is sixty years old, and they have three adult children. The year is 2050, and we join Alisha in the green room backstage at a large auditorium in Mexico City. Alisha is nervously waiting to be honored as one of the most inspirational Hispanic women of the century. Her acceptance speech will be broadcast worldwide to thousands of attendees. Alisha is amazed that the language barrier, which was such a challenge when she was growing up, ceased to be an issue in the mid 2030s. Quantum computing in the Cloud now provides instantaneous, virtually perfect translations of every known language and dialect. Once multilingualism ceased to be a barrier and every person on the planet had access to the Cloud, she was able to easily expand her workforce and business to forty-two countries. Alisha is proud of the result. People in less developed countries are now part of a Cloud-enabled global workforce and marketplace. It is eliminating subsistence poverty by offering first-world opportunities and reducing, if not eradicating, exploitative labor practices.

A staff member pops in to give Alisha the two-minute warning. As she walks to the door, she looks in a mirror for a last-minute hair and make-up check; she's amazed by the perfect fit of her 3-D printed suit designed by her daughter. It's made from recyclable materials that adapt to changing temperatures. For a split second she sees the twenty-six-year-old who in 2020 decided to start her own business rather than accept a partnership at a major accounting firm. She remembers the milestones of the last twenty-five years spent developing her Cloud-based, international, virtual accounting business.

When Alisha was born, her mother worked for the Mexican government in an administrative role. Her father, a US citizen, was a coffee broker who worked mostly in Central and South America.

Alisha was eight years old when her parents divorced, but fortunately both remained involved in her life. Growing up in Mexico, Alisha experienced poverty and social instability firsthand, but she also learned to appreciate her native culture and history and the importance of family.

After graduating high school, she went to university in the United States where she majored in accounting. She was subsequently recruited by a midsize regional CPA firm specializing in providing services to small to midsize businesses and high-net-worth individuals. Four years later, Alisha was recruited by one of the largest US accounting firms. Her work ethic, experience, intelligence, accounting expertise, and language fluency had been noticed. Her bosses told her that she was on the fast track to a partnership.

However, Alisha began to feel that working long hours and having little control over her destiny had limited appeal. Always keeping an eye on the future, she had been watching the growth of artificial intelligence and blockchain technologies. She suspected that they would soon eliminate much of the lower-skilled work currently being done by accounting firms. Alisha was wise enough to recognize that her future accounting career needed to be in specialized niches where she could create high client value using skills and knowledge that are inherently difficult to automate.

Once COVID-19 hit in 2020, Alisha suddenly found herself working from home and utilizing online meetings and videoconferences. She discovered that the new virtual environment helped her overcome her natural shyness. She felt safer speaking up and leading others. Having her ideas rejected, something that had previously terrified her, was somehow easier to handle via video. Even more astonishing was the fact that her increased confidence now extended to face-to-face meetings. It began to dawn on her that working from home, in the Cloud, could become the foundation for the career independence she desired and the family life that was so important to her.

After several months of research, Alisha quit her job to launch an online, Cloud-based, virtual accounting firm. With specialization uppermost in her mind, she targeted small to midsize businesses pioneering emerging technologies in highly regulated industries. Alisha believed that specialized accounting and regulatory compliance in this niche—an area neglected by the major accounting firms—could help clients create major new industries that would be important in the future.

She knew from the outset that she did not want or need a brick-and-mortar office and an organization with full-time employees. The legion of "gig workers" was growing. Her strategy was to build a highly adaptive, motivated team of accountants and auditors who preferred to work as independent contractors. She would motivate and retain the best amongst them by offering minority ownership over time. The company would be virtual, but she would hire independent contractors in each geographical region she was targeting. In this way, she would have feet and eyes on the ground for her clients. By the 2050 conference, her company had over 800 clients across five industries in North America, Europe, and Asia, and 5,000 independent contractors, several hundred of whom were now minority owners.

Alisha seized new opportunities as technology gave rise to them over the years. In the 2030s, her staff was the first to use passenger drones to survey and assess clients' solar farm operations. In the 2030s, they adopted personalized, direct learning and virtual and augmented reality in all training programs and client interactions. These astounding technologies enabled her to rapidly build a highly trained worldwide staff and immerse them in client operations. These and other technologies also eliminated most business travel, lowered operating costs, and contributed in a small way to the reduction of global warming.

As Alisha steps onto the stage, she scans the audience for her husband and children. She sees them, three rows back, applauding. She takes equal pride in her entrepreneurial achievements, her family, and in doing work that makes the world a little better.

After her speech, Alisha is besieged by attendees with questions. It's almost an hour before she reunites with her family. Later, they will fly to Fiji for a vacation on a descendant of one of Elon Musk's Starships: the flight time will be a mere thirty minutes.

~

There's a lot to unpack in Alisha's story, but let's start with the fascinating prediction that Alisha and her family will be able to fly from Mexico City to Fiji in around thirty minutes. It is an example of one of the many imminent innovations that will alter life as we know it. Elon Musk, in a September 2017 presentation to the International Astronautical Congress, stated that, although several years from becoming reality, the technology already existed to fly people from Los Angeles to London in twenty-three minutes and to anywhere in the world in under one hour. Musk predicts that sometime in the near future, his *Falcon 9* reusable rocket will orbit the earth with one hundred passengers at a speed of around 16,000 miles per hour, and then make an upright landing at any one of dozens of launch pads situated twenty miles or so offshore of the world's major coastal cities.* What's more, he predicts the cost will be about the same as an economy airfare ticket today. In case you're thinking this is hyperbole, check out the Starship update video on the SpaceX website.† You'll notice that at the end of 2020, Musk's company began working with the Pentagon on a rocket that can ship cargo and weapons anywhere in the world in less than one hour.‡ Contracts like this will help fund the prototype for the passenger version that Alisha and her family will experience.

The bigger story is that geographical distance will be less relevant in the future. Well before the passenger Starship launches, other transportation options will become available. Think self-driving vehicles, flying Ubers (already being tested), and passenger drones. Alisha's story suggests how travel itself will become less necessary as virtual and augmented reality, along with holographic imaging, better simulate in-person meetings.

Coming back down to earth (pun intended), Alisha's story illustrates that in a digital Cloud-based world, people will be able to create self-determined and self-reliant careers independent of an employer and live and play in a world where geographical distance is almost irrelevant. People will become their own branded product and market their unique ideas, skills, innovations, and creative

---

*   SpaceX, "Starship | Earth to Earth," September 29, 2017, YouTube video, 1:57, https://www.youtube.com/watch?v=zqE-ultsWt0&feature=emb_logo.

†   SpaceX, "Flight Test Starship SN15," video, 16:05, https://www.spacex.com/vehicles/starship/.

‡   Kate Duffy, "The US Military and Elon Musk Are Planning a 7,500-mph Rocket that Can Deliver Weapons Anywhere in the World in an Hour," *Business Insider*, October 9, 2020, https://www.businessinsider.com/musks-spacex-partners-us-military-to-deliver-weapons-by-rockets-2020-10.

content through specialized Cloud-based digital marketplaces.

The story of Susan Merrill in the next section illustrates how much life and career models are changing and explores the role automation will play in our lives.

~

It's 2030. Susan Merrill is forty-two years old and married to Jim who is forty-five, and they have two daughters aged four and seven. The last ten years have been tough. Susan thought she had done everything right. She followed in her parent's footsteps—worked hard in high school, went to college, got a law degree, married, had kids, and became a working mom as a criminal defense lawyer at a large law firm. Jim had a successful corporate law practice with a nationally known firm. Things were rocking in the Merrill household. But outside, things seemed to be falling apart. There were scandals in government, education, and religion. Society was becoming increasingly polarized. And then came the COVID-19 pandemic and everything got much worse—and fast. Many traditional businesses went under, from small services like restaurants and dry cleaners to giant retailers and movie theater chains. Only the tech companies, online retailers, and delivery services seemed to be thriving. Susan had always believed that law was a safe profession. Everyone did, until she was suddenly laid off. Artificial intelligence replaced much of what she did—who would ever have thought that would happen? She tried selling real estate. It worked well around the kids' schedules, but she struggled with the long hours and the absence of a guaranteed income. As a newbie to real estate selling less expensive homes, she noticed how the AI monster had started to replace agents in this sector of the market. She saw the writing on the wall and gave up real estate. Jim kept his job, but AI felt like a piranha—stand still in the river for too long and it'll strip you to the bone. Susan saw their dreams of financial independence and world travel evaporating. Fighting discouragement and questioning her self-worth, Susan decided that the best way to fight AI was to learn what it was and how to use it. She wasn't technologically illiterate, but she was no geek and didn't want to become one. She found her answer in Amy, an electronic assistant that functions as a personalized, voice-activated brain-to-computer interface (BCI). Amy reminded Susan of Siri or Alexa back in the day, but far more advanced and humanized. Susan trained Cloud-based Amy to manage almost every aspect of her family's life, keeping the house at the right temperature, coordinating schedules, ordering food, booking appointments, paying bills, buying gifts, and more.

Susan became so sophisticated in using the Cloud that she converted the family budget and lifestyle to a subscription model instead of relying on owned assets. Unlike ten years ago, here in 2030 Susan and Jim own almost nothing, not even their home. A monthly fee enables them to use whatever they want, whenever they want, including a self-driving Tesla or Uber flying car, streamed entertainment, and delivered meals and groceries. Even their residence is a subscription service, sort of a turbocharged Airbnb or VRBO long-term rental with rights. It was all so easy. She just had to tell Amy what she wanted to do, train her once, and it was done. Susan was proud of what she had achieved. Their family operating costs were lower and could be quickly lowered further if Jim lost his job. In fact, relocating to a new job would be easier without being tied to one home.

Unexpectedly, Jim ripped into her one day. He said that Amy

had become much more than a voice-in-a-box. He felt that Susan related to Amy as a real person, which he thought was creepy and like something out of a science fiction movie. He bitterly complained that "AI Amy," as he called her, oversaw their lives. Jim's complaints caught Susan by surprise, but she realized that he was right. *Damn it!* she thought, *AI got me again. How did I let this happen?* She knew that Amy was software, not human, but she had slipped into treating her as human just as she sometimes did with their pets. Susan also realized that automation had a way of sneaking up on you and making you dependent before you even realized it.

Susan spent the next few days mulling over her life and career. Her parents kept asking when she was going back into corporate life, but she questioned their generation's success model. She also realized that something had always been missing from her careers and jobs. They did not allow her to do what she loved best, and that was to creatively work with people to solve their problems. She'd also been missing significance in her career. She needed to know that her work and life mattered beyond making a living.

Susan decided that her Amy experience was a blessing in disguise. She had successfully used the virtual assistant to reinvent how her family lived. Now she could use it, wisely this time, to reset her career direction. Instead of delegating to Amy, she began to think of "her" as an extension of her own power, a collaborating, expanded capability. Energized, she initiated a research project. She instructed Amy to search and analyze "emerging independent lifestyles," "emerging trends in education," and "successful female-owned Cloud-based businesses." That's when she came across Alisha's female-entrepreneur-rising-star story. From there it wasn't a huge leap to design and launch a new virtual dispute resolution business aimed at helping people avoid litigation costs.

Her business-in-the-Cloud targeted countries with notoriously poor legal systems and areas of the law where mediation was a better solution than the increasingly ineffective judicial system.

Now, in 2030, the business is going well. Jim plans to quit his job and work with Susan as soon as it becomes financially viable. They plan to work from anywhere (WFA) through the Cloud while traveling the world instead of waiting for retirement. They will use their subscription services to live for a year in each of the countries Susan has targeted for expansion while building a network of high-quality, local dispute resolution experts. They will homeschool their kids with personalized remote learning. Owning nothing means moving is easy—wherever they go, everything will be set up for them, just like at home. Although "home" itself is becoming a relative thing. And it all began with Amy. Susan and Jim's lifestyle choice will not be for everyone, but many will try it, at least for part of their lives.

Susan's story demonstrates what could happen when people utilize the power of the Cloud with nothing more than their voice. Used wisely, seamlessly integrated automation empowers people. Susan's effective use of Amy made a new career and lifestyle possible for her and Jim. Voice-activated, Cloud-enabled automated assistants will emerge and become more powerful over the next thirty years. They may be contained in watches, eyeglasses, bracelets, necklaces, or other small wearables that we cannot yet imagine. The next generation of brain-to-computer interface (BCI), following voice activation, will be thought. Imagine all the power and functionality of today's smartphone in a tiny chip. Now multiply that power by a big number, perhaps ten or ten squared. Then imagine that the chip directly connects to your brain by being implanted, or non-invasively attached using something like a wearable patch or earpiece. With thought activation, you won't need to type, speak, or look at a monitor to engage the virtually unlimited power and information available in the Cloud. You would no longer need to text or call anyone—simply think and it happens.

The importance of seamless connection to automation, activated solely by voice and thought, cannot be overstated. Any person, anywhere in the world—24/7/365—will be able to effortlessly engage advanced Cloud-based computing power and unlimited information and services. It will include people living at subsistence level, children, people with disabilities, and the illiterate. No knowledge of computing will be required. You will simply speak or think what it is that you want to do. Today, dozens of companies are developing brain-activated brain-to-computer interfaces (BCIs) to address motor skill disabilities and diseases like Parkinson's. Within thirty years, robotics and exoskeletons powered by these innovations will enable the elderly and disabled to overcome many of their body and mind limitations.

Our first two futuristic examples were about highly motivated business entrepreneurs, but not everyone is cut out for running a business. Corporations will still need employees, but make no mistake, the corporate career landscape is going to change dramatically.

~

In the following story that takes place in 2045, we follow Greg and Linda, ages thirty-nine and forty, respectively, who were teenagers at the time of the global pandemic of 2020. They were both born in Columbus, Ohio. Their parents were middle-class, educated, and held down good jobs until the pandemic hit the world like a tornado, ripping apart everything they felt was secure and safe. By the time Greg and Linda left college, their expectations about life and career were very different from their parents. Linda was an avid reader of Peter Diamandis and had read his books, *Abundance: The Future Is Better Than You Think*; *BOLD: How to Go Big, Create Wealth and Impact the World*; and *The Future Is Faster Than You Think*. By the time she was eighteen, she already knew that she needed to think differently than her parents about her life and career—very differently. Before she went to college, she was determined to reach her potential by keeping herself ahead of the trends in a fast-changing world.

Greg and Linda met at a cocktail party during a lively discussion about current events and the directions of things in the future. Greg was amazed by Linda's depth of knowledge about where the world was headed. Linda later encouraged him to read futurists Alvin Toffler and Ray Kurzweil. Kurzweil was a futurist, inventor, and entrepreneur. He had a long history of making accurate predictions about the world that Greg and Linda would be navigating over the next several decades.

Fast forward to 2045, and they are sitting next to their pool in Thailand. They have both done exceedingly well for themselves in their respective careers. Greg is a data engineer for an American AI company, and Linda is a fashion designer for a leading online clothing mega-company.

In 2020, when they were fourteen and fifteen years old, there were 125 million people who were age eighty and older; in 2045 there are over 400 million.* The reality is that in 2045, Greg and Linda are classified as young people—a classification that will continue until they are over sixty-five when they will gracefully move into middle age. Since they were teenagers, artificial intelligence has fueled extraordinary breakthroughs in the treatment of most of the diseases that killed previous generations. Wearables track health 24/7/365. 3-D printed body parts and organs such as kidneys and livers are commonplace. We will delve deeper into the importance of health when dealing with the New Realities in a later chapter.

Greg and Linda are financially sound and have a comfortable lifestyle. They have a well-planned, liquid, flexible, and professionally managed investment portfolio that includes a healthy contingency fund. The latter is important because they are constantly presented with investment opportunities that could help fund their long lives and extended retirements.

However, they don't own a home, a car, or much of anything else. Asset accumulation, as it is with many professionals in 2045, is low on their priority list. They do however pay a substantial monthly "living free" subscription that provides them with luxury accommodation almost anywhere they wish to visit in the world. It's like owning homes in dozens of countries. No matter where and when they arrive, a vehicle is available as part of the subscription. They also subscribe to a service that allows them to borrow physical art from local museums and art galleries or rent virtual art while using the homes. In this way they can enjoy and experience different cultural experiences wherever they happen to be at any given time.

With no encumbrances—and jobs that allow them to work

---

* "Ageing and Health," World Health Organization, February 5, 2018, https://www.who.int/news-room/fact-sheets/detail/ageing-and-health.

remotely—they travel the world as part of their life and careers. They have been in Thailand for six months. The Thai cities have amazingly fast Internet speeds, vibrant and exciting expat communities, and low-price accommodations. Their subscription has allowed them to live close to downtown in a house that they could only dream of experiencing in New York, London, or San Francisco. In a few weeks, they will move to a smaller, but no less stunning, villa in Todi, Italy. After that, they plan to "settle down" and spend a few years near family in the United States before choosing the next country they will call home for as long as they feel inspired.

Our next story highlights the importance of being prepared for the not-too-distant opportunity to extend one's career and life significantly. It is human nature to think short-term, but this story demonstrates how thinking ahead and acting now may literally save your life.

～

Peter is eighty and his partner, Ed, is sixty-five. The year is 2040. Both are fit and healthy. Peter is a particularly youthful and vibrant octogenarian. We join them in Peter's doctor's office. Dr. Bill Myers has been Peter's physician for more than twenty years and is a specialist in the emerging field of healthy longevity. His office had called Peter to set up an appointment to discuss a medical breakthrough. After exchanging pleasantries, Dr. Myers informs Peter and Ed of a recent longevity discovery that could extend Peter's life by twenty to thirty healthy and active years. He cautions, however, that there are two preconditions for the treatment. First, candidates must be in extraordinarily good health. Thankfully, Peter qualifies. The second is that life-extending therapies are considered cosmetic and experimental by government and private health plans, therefore insurance will not cover the treatment cost of $250,000. Peter takes a deep breath and looks at his younger husband. "A quarter of a million dollars. *Wow!*" Ed looks at Peter and almost imperceptibly nods, a smile gently spreading from his eyes to his mouth. Peter turns to his doctor: "Schedule the procedure, Bill."

This happy outcome was not by chance. It was the result of Peter's decisions and actions beginning twenty years earlier. In 2020, Peter was enjoying a successful career as a management consultant in a sizable Boston-based firm. He and Ed lived a good life, and they never worried much about tomorrow. Most of their retirement assets, despite warnings from their financial advisor, were invested in their home. Peter's health was average at best. His lifestyle was short on exercise and sleep and long on stress, wine, and dessert.

Then the COVID-19 pandemic hit and changed Peter's life. He was furloughed without pay and had to take on a second mortgage to pay the bills. Two senior partners at his firm died from the virus, both of whom were in their sixties with multiple pre-existing health problems. It was a wakeup call. Peter made three commitments to himself and to Ed. He would achieve the best possible health status and level of fitness for his age. He would work past his previously planned retirement age, if possible, and diversify their sources of income so that they would not be dependent upon his employer. They would lower their household costs to create a liquid, readily available cash reserve for emergencies.

While the pandemic was still raging, Peter began a regimen of daily exercise. He noticed many other people doing the same. He dropped his daily ritual of at least two glasses of wine, replacing it with an occasional half-glass treat. He found that his sleep immediately improved. As the pandemic waned, Peter joined a gym for the first time and began a formal fitness and health program. He also sought out a healthy longevity advisor and signed up with Dr. Myers.

Peter returned to his job after the pandemic; fitter, more alert, full of energy, and eager to work. As he approached his firm's traditional retirement age a few years later, he was pleased when they asked him to stay on. Years of falling birth rates had resulted in a vastly smaller pool of talented, qualified, and experienced consultants. Compounding this was the fact that the consulting industry itself had changed. Clients were no longer willing to pay high fees for young, inexperienced consultants with a senior partner as the titular project leader on their account. They demanded the wisdom, maturity, and relationship skills of senior partners on their projects. Peter was the perfect fit for clients because of his experience, energy, good health, and mental acuity.

Peter delivered on his financial planning promises. He and Ed downsized to a home with lower maintenance costs to increase retirement assets and build an emergency cash reserve. They utilized the extra years of Peter's consulting career to diversify their income. Starting with a single rental property, they built a portfolio of long-term and vacation rental homes, and a boutique residence hotel for the growing "residence as a service" market.

Imagine living to 100 or 120 years old and being as vibrant as today's fit seventy-year-olds. Imagine using your creativity and wisdom in a career that lasts thirty, fifty, or seventy years to reach your potential and make a difference. Maybe you could be part of solving a significant world problem, solving a series of smaller ones, or making life better in other less notable ways. Maybe you could play a role in opening new frontiers under the sea or in space. Not everyone will work on expanding frontiers or life-changing innovations. However, more and more people will have unprecedented opportunity in a Cloud-based digital economy, as our next story illustrates.

~

Gary is in his early fifties; the year is 2040. He owns and operates a custom woodworking factory making rustic, country-style furniture and sells his products online. Gary began in 2021 by building and selling knick-knacks, signs, and other items made from wood for the home while working a full-time job as a local delivery service driver. Today, he has an extensive e-commerce website and 3-D prints models of his custom designs for his high-end clients. In addition, he uses a virtual reality app to provide his clients an immersive experience of how the furniture will look in their home. They can smell the fresh-cut wood and stain, they can sit at a dining room table, or open the doors of a sideboard to look inside. His clients "see" and experience the furniture in their home before he cuts the first piece of wood.

With only a high school education, Gary had once believed that he was destined for a career in manual labor working for others. Twenty years later, his company has annual revenues of $40 million in 2021 dollars. He was an early adopter of reaching for his potential when most of his peers continued to follow career models that were becoming outdated or ineffective. Gary's career might once have been considered craftsman work, or even blue collar, but became "light-blue" through integrating the craft with leading-edge technology.

~

The twentieth century career model of lifetime employment with one or two companies or in one career with retirement at sixty-five is largely obsolete and will become entirely a thing of the past in the next thirty years. The twenty-first century career model involves a longer, healthier work life with multiple careers, jobs, and employers, or an independent career as an entrepreneur or gig worker, or perhaps a combination. There will be multiple transition and retraining periods between jobs and careers, often without income. A liquid financial reserve will be extremely valuable to bridge career and job gaps, maintain your health, and supplement retirement savings. The COVID-19 pandemic demonstrated the importance of this new twenty-first-century model. It would be wise for those under the age of fifty-five to adopt it, as Peter did in our story earlier.

The innovations discussed in this chapter will make people superhuman by historical standards. To the degree that people become dependent upon those innovations, they will become cyborgs. This New Reality may sound like a mind-boggling opportunity or a terrifying nightmare depending on whether you are an optimist or dystopian. There are arguments to support both ways of thinking that will be addressed in the following chapters.

Regardless of your feelings, these innovations are coming, and you will have to decide if and how you will assimilate them into your life. Susan's story illustrates that ignoring or fighting automation is likely to limit or destroy a career. But Susan's experience also demonstrates that, as the human in the relationship, you must remain in charge. Many people today have difficulty walking away from their technology and engaging in the life going on around them. We can see it with smartphones. Imagine how much more difficult it will be to turn off your voice-activated automated assistant with human qualities or to disengage your brain from a thought-based Cloud brain-to-computer interface (BCI) that allows you to function at near-genius levels. Another important consideration will be security and privacy. The more technology-dependent we become, the more transparent we are to anyone who wants to know about us. Everyone will need to assess the adequacy of security and privacy before integrating new technology into their lives.

Alisha, Peter, Susan, Greg and Linda, and Gary's stories provide glimpses into the future of fast-paced technological, cultural, and societal change. In the next chapter, we will explore the anatomy and mechanics of change so that you can prepare and navigate the next thirty years without fear or confusion. Later, we'll propose how to master skills that can make your career and life exceptional and open doors to your potential.

PART I

The Forces
of Change

"THROUGH THE LOOKING-GLASS" MAY HAVE left
you feeling a little anxious about the future, but equally captivat-
ed and curious by its wonders. Psychologist Dr. Lisa Colangelo
Fischer says such feelings are predictable. Paraphrasing a quote of
unknown origins often attributed incorrectly to Lao Tzu, Warren
Buffett, and Brazilian Junio Bretas, Dr. Fischer reminds us: "It is
human nature to focus on today. History can seem baffling or irrelevant and the unknown future makes us anxious. We immerse
ourselves in the small things of the present that make us feel safe."
Putting one's head in the sand has never been a good idea, but
it can be perilous when times are changing rapidly. Author and

futurist Alvin Toffler said, "You've got to think about big things while you're doing small things, so that all the small things go in the right direction." We can understand the *direction* of change and its forces and patterns thanks to the work of brilliant futurists and scientists past and present. With that understanding, we can grasp how change will affect our careers, lives, and families. We can adapt earlier and easier to avoid the pitfalls and capitalize on the opportunities that periods of great change inevitably offer. Having a broad perspective of the changes underway and a plan to adapt will be a distinct advantage in the coming decades.

Providing an understanding of the direction and forces of change, and the New Reality that will result, is the purpose of the next four chapters. Note that what follows in these four chapters is *not what I wish to happen*, but what I believe is *likely to happen* based on the references in the bibliography, innovation progress, and my own observations. I hope to stimulate your interest to pay close attention as changes unfold in the next three decades so that you are prepared to adapt quickly.

Consider the value of being able to identify careers that are likely to grow and flourish, rather than decline or potentially be eliminated by automation. It's not just careers that are volatile; your retirement will likely be delayed until much later in life—and it won't be your father's retirement. Understanding how to plan for and afford a "New Reality retirement" with all its nuances will be vitally important. Change-induced instability and volatility will be most evident in the number of companies and products that will face disruption. Predicting this disruption will become a valuable skill. It will allow you to change careers or take advantage of an entrepreneurial opportunity before disruption occurs. Taking this one step further, what if you could foresee future disruptions in government, religion, education, media, and communications and assess in advance their effect on you and your family? What if you could understand, in real time, how innovations are combining with other forces of change to create the "New Reality" that is your life?

CHAPTER 1

Innovation-Resistance

> "Innovation is the ability to see change as opportunity, not a threat." —Steve Jobs

> "Difficulties are meant to rouse, not discourage. The human spirit is to grow strong by conflict." —William Ellery Channing

YOU ARE NOT ALONE IF you think things are changing faster than they did when you were younger, or that today's world is becoming increasingly turbulent and volatile.

> Things are changing faster than ever, and the pace of change will continue to accelerate, making life far more complex than in your parent's and grandparents' time.

Yet, our goals are much the same: pursue an education, find a good job, perhaps get married and have a family, live in a comfortable home, travel, have time for ourselves, and enjoy a comfortable retirement or less harried pace late in life.

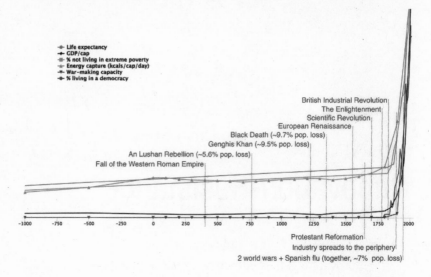

Fig. 1 Acceleration of change (Graph by Luke Muehlhauser. lukemuehlhauser.com.)*

Since the First Industrial Revolution of the 1760s, the pace of change has been accelerating. Pace of change can be measured by the time between events that significantly alter everyday life and living conditions. This chart, created by Luke Muehlhauser, illustrates this point.

But why is change accelerating and to what end? Let's examine the evolutionary forces that are remaking the world as we know it, and the "New Reality" in which we will live.

---

*   Luke Muehlhauser, "How Big a Deal Was the Industrial Revolution?," lukemuehlhauser.com, accessed January 17, 2022, http://lukemuehl-hauser.com/industrial-revolution/.

For example, think about the weather. Forces such as heat, high-altitude winds, the oceans, the seasons, and other factors combine to produce rain, snow, or sunshine. Most of us think little about these forces, we simply check our weather app to see if it's going to be sunny today or whether we'll need an umbrella. Early civilizations believed that weather ruining their crops was a result of the gods being angry with them. They might even have felt the need to offer a human sacrifice. Today, we accept the weather as the result of natural forces. By the same token, we should accept the forces of change that converge to produce the New Realities of life as natural, not to be feared but to be understood.

The first change force we'll examine is *innovation*, along with its mirror image, *resistance*.

> It's the tug of war between innovation and resistance that governs the pace of change and the volatility we experience.

If we look back at Muehlhauser's chart (see Figure 1), the pace of change was very slow at first. It took our hunter-gatherer ancestors millions of years to evolve and leave Africa. It was a mere 10,000 or so years ago that human beings began to settle as farmers. Even following this evolutionary milestone, life-altering change remained incremental, perhaps only one or two in a century or a generation, until the First Industrial Revolution. Then the age of innovation truly began, and the pace of change accelerated and has continued to accelerate. Innovations don't create an isolated effect. Each innovation alters the lives and conditions it touches. Each innovation in turn increases the capacity to innovate in other ways. Thus, the original and subsequent innovations, and all that they in turn touch, accelerate the pace of change. That acceleration continues until today where the world, and especially the United States, has developed a culture of innovation. As the arc of history shows, and our personal experiences validate, the well-being of the world and humanity has improved in terms of health, increased lifespan, reduced poverty, and technological innovation. And, although the media sometimes seems determined to deny it, we live in a better, safer, more abundant world compared to any other time in history. However, as we will explore throughout this book, humanity is far from reaching its potential.

Resistance

Many of us get excited about innovation and are eager to experience the latest and greatest technology the world has to offer. However, we usually fail to consider the resistance and possible adversity that accompanies innovation before we adopt it. As every action has an equal and opposite reaction, so every innovation brings resistance in the form of its unexpected complexities, barriers to use, and unintended consequences, including making life busier or more complicated instead of simpler as anticipated.

Most of us have experienced the phenomenon of resistance as new systems or devices at work or home surprise us with unanticipated advantages and disadvantages. Sometimes they have a significant negative effect on our lives. For example, the Internet rivals the printing press for having an overall positive effect on the world. But it has also facilitated an unanticipated, unprecedented assault on privacy, security, and civil liberties that is becoming increasingly problematic. The smartphone is a miracle of technology that has become a staple of everyday living, but many people have become its slave. On-demand ride services such as Uber and Lyft won immediate consumer acceptance but met fierce resistance from taxi services; attempts to unionize drivers; and isolated, catastrophic events arising from inadequate vetting of drivers and riders. Unexpected consequences and resistance can often be severe enough to delay innovation for decades or cause potentially valuable innovations to be abandoned altogether. Resistance is why the timing of change is harder to predict than whether it will occur at all. Science fiction books and movies and futurists often predict the future fairly accurately but get the timing wrong.

Resistance, however, should not be seen as a malicious force, but as a natural, beneficial function. Like adversity in our personal lives, resistance improves the end result by serving as a quality control check. It puts the brake on things until we get them right and at the same time regulates the pace of change. Too much resistance delays progress and evolution. Too little causes excessive, and potentially dangerous, unintended consequences because insufficient time has been dedicated to working out the bugs or to find ways to safely integrate the innovation into our lives. If we accept that every innovation will have unintended consequences, we can speculate on what they might be and perhaps learn to anticipate and avoid some of the negative effects. Perhaps in the future, probability models will assist us before we adopt an innovation in our everyday lives. We should always ask ourselves, *What could go wrong?* It will be important in the future to be wiser and more deliberate when adopting innovations as they come at us at a more rapid pace.

Innovation

Revisit the previous Muehlhauser chart (Figure 1), but now with resistance in mind. If the pace of change is accelerating at its current rate *despite* resistance, what does that tell us about the level of innovation and our desire to adopt it? Humans are infinitely creative, and necessity is truly the mother of invention. Each time humanity faces a challenge, our collective creative spirit rises to resolve it, often solving problems that we never set out to solve. Unfortunately, along the way, we often create new unanticipated problems.

A story that illustrates this point well comes from the book *Super-Freakonomics* by Steven D. Levitt and Stephen J. Dubner. The urban environment in the 1890s became overwhelmed by mountains of manure created by horses and mules, which was the usual mode of transportation at the time. Not anticipating the invention of the internal combustion engine, one "expert" predicted that manure would reach the fourth floor of Manhattan buildings within a decade. From this perspective, the appearance of the automobile a few years later was seen as environmentally friendly. Today we see the automobile's combustion engine in a different light. The reasons that we are predisposed to think negatively are discussed in Chapter 4.

Somewhat ridiculous, panicky prognostications are as common today as they were in the early twentieth century—as a few minutes on a news feed or social media will illustrate. Future generations will view today's fears of disaster to be as silly as the prospect of forty-eight-foot-deep horse manure in Manhattan streets is to us today. As Levitt and Dubner say in the book, "Technological fixes are often far simpler, and therefore cheaper, than the doomsayers could have imagined."

It would be easy to treat this story as simply a cute anecdote, except that we continue to fall for the same doomsday line repeatedly. The moral of the story is that dire prognostications, by their nature, will almost always be overstated, and we generally fail to anticipate the innovations we cannot yet envision.

A series of mind-blowing innovations, including those in the following list, will arrive in the next three decades. They have the potential, individually and collectively, to address today's dire catastrophic predictions but each will also bring resistance in the form of new problems and unintended consequences.

"Seamless" brain-to-computer interfaces (BCIs) will vastly *extend* human capacity easily and naturally and in parallel *expand* opportunity worldwide as never before. Seamless automation means that we will become incrementally powerful without feeling a dramatic change. Seamless automation will feel more like using contact lenses, or a turbocharged cup of coffee than a separate device or new skill to learn.

Artificial intelligence (AI) and robotics will expand human power. They will free humans from repetitious, dangerous, and physical work better carried out by machines and enable human work that has much higher value. AI will slash the cost of automation and the time from identifying a problem or opportunity to making it a reality. That will hyper-accelerate the pace of change.

The Cloud will become as ubiquitous and essential as electricity. It will encompass almost every aspect of work and life. The amazing potential of the Internet of Things (IoT) will be unleashed. Sensors and devices that we use in our everyday lives will not only provide vital information but will personalize the information and responses to it based upon predefined preferences without manual intervention. This new vision of the Cloud or the next version of the Internet is often referred to as the *metaverse.* On October 28, 2021, the giant social media company Facebook announced its new strategic direction and name—Meta—in a letter from its founder, Mark Zuckerberg. The letter is significant because it commits one of the world's largest tech companies to an astonishing vision in which experiences are shared in an alternative reality as if we are there in person. I use the term "Cloud" throughout this book because more readers are familiar with it and can relate to the experience. However, the Cloud of the future is not limited to what you use today. It will incorporate the envisioned functionality of the Internet of Things (IoT) and the Metaverse.

Quantum computers promise an increase in computer power that is hard for us to wrap our heads around—reportedly tens of thousands (if not millions) of times faster than conventional computers. Such power will expand the limits of what we can model and comprehend as well as the problems that we will be able to solve. Consider the implications of putting that amount of computing power into solving environmental issues, the aging process, healthcare, alternatives and defenses to war, and many other humanitarian challenges. However, like any innovation, quantum computing will bring resistance and adversity. For example, a quantum computer would likely overpower the security features of traditional computers, meaning that a defense against misuse must move in parallel to adoption worldwide.

Mobility innovations will open the door to faster and safer travel, including space travel. However, traveling will become optional in most cases as AI and virtual reality and augmented reality and holographic technology make experiences increasingly realistic.

Medicine and automation will converge to create the science of healthy longevity, extending lifespans and our quality of life.

The positive benefits of these innovations come with greater responsibility individually and collectively. We must do everything in our power to advocate and support innovation to solve our most pressing problems. Simultaneously, we must improve our ability to predict and offset unintended and unanticipated consequences.

Volatility

Volatility is both a direct and indirect result of innovation. The faster the pace of innovation and change, the greater the resistance.

> The interaction between innovation that is trying to advance life and resistance pushing back creates volatility.

As the pace of innovation and change accelerates, the span of effect broadens and converges with other changes, obscuring cause and effect. Volatility occurs and events appear to come out of nowhere. Author Nassim Nicholas Taleb called such events "Black Swans." In his book, *The Black Swan,* he explains what he calls "the impact of the highly improbable." Volatility that seems to have no directly relatable cause unnerves us far more than when we understand its origin.

A recent example was the COVID-19 pandemic. Public health officials had warned of the potential danger of pandemics for years.

Several epidemics were pandemic near misses. Yet we all went on with life, ignoring the warnings as if a pandemic could not and would not happen. Governments and organizations, and we as individuals and families, failed to develop contingency plans. In parallel, low-cost airfare and tourism increased dramatically with little thought to how quickly a malicious, lethal virus could use this easy conduit to spread globally.

As will be discussed later, politics, news and entertainment media, and a huge amount of marketing use fear to manipulate us. When so many things in our lives use fear to get our attention, when there are so many voices crying wolf, it causes us individually and collectively to miss the smaller, highly important matters that should in fact be our highest priority.

There will undoubtedly be more Black Swan events. They will seemingly come out of nowhere as the pace of innovation, change, and resistance accelerates. Planning for them, expecting them, and doing everything possible to mitigate their effect will be part of our life going forward.

CHAPTER 2
Democratization

> "You know, I believe that technology is the great leveler. Technology permits anybody to play. And in some ways, I think technology—it's not only a great tool for democratization, but it's a great tool for eliminating prejudice and advancing meritocracies." —Carly Fiorina

THE WORD *DEMOCRATIZATION* IS UNFAMILIAR to many people and is often incorrectly associated with politics. Think of democratization as an evolutionary force that has been with us since the beginning of time. It empowers us to assume greater responsibility for ourselves and to play a larger role in the world around us.

Democratization and innovation are interrelated in a chicken-and-egg enigma. The drive to democratize sparks innovation, and innovation enables democratization—which comes first is irrelevant for our purposes. We need only recognize that democratization is occurring all around us at a much faster pace today, reshaping our lives and careers in ways previously unimaginable.

Individuals adapt faster than organizations to democratization, innovation, and accelerating change. Individuals can make decisions and change their lives overnight because they have few, if any, other people they need to consult. However, organizations have many conflicting constituencies to convince and processes to adapt—both of which can take months, years, or even decades. An organization's owners, leaders, and employees may be vested in preserving the status quo, causing them to deny or resist change even if inevitable. Organizations that perform poorly or adapt too slowly are disrupted when their customers choose alternatives to their goods and services. If they fail to adapt quickly enough or effectively, they de facto deinstitutionalize by becoming less relevant, or they disappear altogether.

> You will not always experience democratization as a benevolent force.

For example, your employer or favorite store may have closed because of competition from online shopping. You may be frustrated by the con artists, propagandists, and predators populating your email inbox. You may be excited to offer your own product or service on the Cloud only to discover the challenge of getting it noticed because other democratized people are doing the same thing. Expanded choice is wonderful as a consumer but finding what you want among so many choices can be difficult and confusing. Or, on a scarier note, most of those new geniuses empowered by seamless voice or thought-activated brain-to-computer interfaces (BCIs) will become supermen and superwomen, but a few will undoubtedly attempt to become supervillains like Superman's evil archenemy Lex Luthor.

Democratization occurs progressively over time, as you will see in the examples to follow. As previously noted, your ancestors adapted to a handful of innovations over decades or centuries. You, on the other hand, will need to adapt to thousands of innovations in a fraction of that time. This compression of time is an acceleration of evolution, less time to understand and make adaptive decisions. It requires advance preparation rather than reacting as change unfolds as we have done historically. Adapting and prospering is what this book is all about. Creating a twenty-first century mindset, skills, and goals are addressed in the chapters ahead.

Let's examine the terms, examples, and characteristics of democratization and deinstitutionalization to illustrate how they affect your career and life so that you can use them to your advantage.

Definitions

The following definitions drawn from numerous sources are simplified and tailored for the theme of this book.

> **Democratization** is "the action of making something (and in total everything) available or accessible to everyone." Keep in mind that the definition is *not* "the action of making something available or accessible to nice people" but to "to *everyone*."

> **Deinstitutionalization** in this context means "the process of abolishing a practice that has been considered a norm."

As democratization and innovation make goods or services available to everyone, existing organizations either adapt or lose their customers to organizations that have adapted or are themselves innovators. Adaptive organizations win; organizations that do not adapt deinstitutionalize.

Disruption occurs when the effects of democratization and innovation become apparent to the organization's stakeholders such as customers or beneficiaries, management and employees, suppliers, investors, and ultimately the general public. Customers are often the first to detect and react to an organization's products and services not keeping up with the demands of democratization and innovation. For example, customers of companies like Sears and Blockbuster were changing buying habits faster than Sears and Blockbuster management and investors detected and reacted. Interestingly, in their own way, both companies had once been democratizing forces themselves but failed to adequately react to competitors Amazon and Netflix. The speed and effectiveness of an organization's response to disruption determines whether it becomes deinstitutionalized, and ultimately determines its survival. It will be essential for you to spot organizational disruption so that you can benefit from it rather than have your career and life equally disrupted.

Organization is used generically throughout this book to refer to all types of organizations, regardless of legal structure or purpose, and includes large and small corporations, partnerships, not-for-profit social institutions, charities, etc. Every product, service, industry, organization, and institution will be reinvented by democratization, innovation, and deinstitutionalization over the next three decades. That means that the effect of these forces on your life and career is inescapable. It will either create havoc or opportunity for you based on your level of preparedness.

Institutions are a special type of organization with social contracts enshrined in law or practice. Historically, they have provided stability, wisdom, and guidance to society. Institutions include all levels and branches of government, organized religions, public and private education, the news media, and others.

Consumer is used generically throughout this book to refer to the people organizations serve. In this context, it includes customers, beneficiaries, constituents, donors, voters, parishioners, parents, and students.

Examples of Democratization, Disruption, and Deinstitutionalization at Work

Democratization of information and knowledge

In today's world, we take for granted that we can find information on any topic we choose because most of the world can read independently. However, for most of human history people relied on other people for information. Until democratized by the printing press in 1440, information and knowledge were mostly available only to the wealthy, the powerful, and the clergy. In recent years, the Internet has had an even more profound democratizing effect than Johannes Gutenberg's press. The accumulated knowledge of humankind is now accessible to everyone with Internet access—and soon that will mean everyone on the planet.

Democratization of creative content

Making movies was difficult, extremely costly, and highly controlled until recently. Movie theater operators and distributors decided which movies would be shown, and movie studios decided which movies were made. Access to creative talent and movie studios was controlled by powerful talent agencies. A similar intermediary structure, or hierarchy, existed in book publishing and in music. Today, we are witnessing their deinstitutionalization. Creative artists have many new alternative paths to consumers that bypass the traditional middlemen and expand consumer choice. Examples include Netflix for movies; Amazon for movies, books, and music; TikTok for comedy, dance, education, and absurd human behavior of all sorts; YouTube for creative performers, including talented felines and how-to videos for just about anything, and TED Talks for thought leaders.

Democratization of goods

Prior to the eighteenth century, most goods were custom-made for wealthy people and homemade for the general population. The First Industrial Revolution in 1760 standardized products and processes, enabled mass production, and democratized goods and services. In 1892, Sears and Roebuck offered an unprecedented variety of goods through their catalogs and shipped them direct to consumers. Mid-twentieth-century big-box retailers dented the catalog business and replaced many local specialized small retailers. In 1994, Amazon began to replace printed catalogs with online ordering, fast home delivery, and easy returns. Sears and

Montgomery Ward, big-box retailers, and tens of thousands of small retailers were forced to become online retailers to compete. Many could not compete and became obsolete. In a 2020 report, Credit Suisse estimated that 25 percent of shopping malls would close within five years. Ultimately, 3-D printing may replace many of today's online and storefront retailers and their intermediaries for many types of goods. Consumers will be able to design and manufacture high-quality customized goods at home or using a local 3-D printing supplier.

Democratization of war

War has been described as "the act of pursuing political objectives through the use of violence." Historically, that meant armies of young men slaughtering each other on a battlefield. That ancient form of warfare effectively ended with the 1993 Gulf War when Saddam Hussein's Republican Guard, comprised of elite soldiers from the Iraqi army, was obliterated in hours by specialized aircraft. Today, most wars employ targeted air battles, drones, combat between financed proxies, cat and mouse operations between terrorist cells and highly trained special forces, and political and cyberwarfare. In recent years, an unfortunate democratization created lone-wolf terrorists who attack civilians or conduct cyberwarfare without nation-state sponsorship. Today, anyone can wage war on anyone or anything, any place at any time. Anyone can be a war victim. Cyberwarfare will likely become more pervasive because it is available to more perpetrators and can achieve political ends without costly death and destruction. But for the first time in history, we are no longer killing hundreds of thousands of nineteen-year-olds every year on battlefields. Good thing. We are going to need them as you will see in the next chapter.

Democratization of society's institutions

Institutional disruption is painful and surreal to watch but is inevitable and universal. Ask yourself, is any societal institution *more* effective, stable, or unmarred by scandal than ten years ago? Are any of their customers or stakeholders happier? Government, religion, education, the news media, and just about every other institution seem to be on slippery ground. Why is this? What is happening?

Institutions, in general, are essential to society. However, they often wield undue power over those whom they serve and represent. They are legally privileged, run by an elite, serve conflicting constituencies, are monopolistic or near-so, and function as intermediaries between us and something we need. They resist allowing greater consumer power, choice, or competition, and they are difficult to moderate or change.

Nevertheless, some institutions, maybe the majority, will adapt and emerge stronger in democratized forms. The arc of history suggests that institutional disruption, polarization, and conflict heralds the birth of something new and better, but the transition could be long and difficult. Specific examples of institutional challenges follow. They could make the next three decades feel more like society disintegrating than a transition to something better. Consider how you will personally navigate these times without fear, seize opportunity, and aid your loved ones and coworkers to do so as well.

Nation-states will need to transfer more power to the governed, decentralize control and administration, replace an insulated self-serving elite with collaborative people and processes, and deliver more value at lower cost with fewer intermediary steps. Resistance to these changes will be intense from the elite and those who benefit from the status quo. An earlier, similar transition occurred during the eighteenth and nineteenth centuries as kings gave way to parliamentary democracies. Worldwide we are witnessing systemic polarization between branches and levels of government, political ideologies, and between government and the governed. Despots and elitists at home and abroad are democratization's anachronisms. They use propaganda, corruption, and violence to promote conflict between polarized elements and distract the governed. They will sooner or later fade. Representative democracy is the most responsive and decentralized form of government, but it has powerful interests that resist democratization's drive to devolve power to consumers (voters). Unfortunately, a vision is yet to emerge for a democratized representative democracy (Democracy 2.0). If you are under age fifty-five you will almost surely be part of the debate to redefine government's role in a democratizing digital world. It will undoubtedly arrive at some point, and when it does, it will reflect the characteristics of democratization outlined in the following sections.

Religious institutions struggle to attract followers while maintaining traditions and orthodoxy that younger generations seem to find less relevant. There is a trend toward people defining themselves as "spiritual but not religious." A growing body of young people are becoming anti-religious. Some are influenced by hostile political ideologies attempting to substitute political ideology for religion. Many young people born in the twenty-first century have no personal experience with organized religion except through news and entertainment media that in general seems hostile to religion. Many of these young people do not associate religion with people's innate spiritual psychological function, but instead with scandal, war, fanaticism, and declining relevance to daily life. In general, many people seem to seek a more tolerant, inclusive ethos and community. To prosper, religions will need to reverse this decline far more effectively than they have thus far. They will need to eliminate all but essential intermediary structures and processes. As one member of the faithful said about his church's huge infrastructure, "I ask myself if we really need all this to introduce people to God."

Educational institutions at all levels have powerful competing constituencies whose interests often supersede, or distort, their mission to serve students and parents. Examples include the interests of faculty, unions, alumni, state and federal regulators, legislators and elected officials, student debt holders, and special-interest parent and student organizations. Some of these constituencies are instruments of political ideology with influence over the curriculum that both parents and students find undesirable. The "sage-on-a-stage," one-size-fits-all instructional method is millennia old and largely outdated. Educators complain of ever-rising administrative costs and burden with little value. Parents, students, and taxpayers complain of rising costs with little to no improvement in the level of education delivered. Curricula needs updating to keep pace, or even in touch, with a digital Cloud-based world. On the horizon is a massive disruption to education as we know it. It will occur as parents, students, and even employers question rising costs, declining value, and the politization of education. A democratized, personalized, low-cost instructional system, available to everyone anywhere in the world, is almost sure to emerge from technology innovations like worldwide broadband, virtual and augmented reality, and educational innovations like micro-schools. Educational institutions will either lead or be disrupted by the transition.

News media news was once among the most venerated of institutions, trusted to provide balanced, objective, fact-based reports independent of political, business, or other powerful interests. An independent media is essential to democracy and is always tyranny's first casualty. Today, news media is viewed by most consumers, and many medical and mental health professionals, as unhealthy, unreliable, negative, sensationalist, biased, or propagandist. Yet news business models remain stubbornly unchanged and unresponsive even as their market share, respect, and relevance to society are in freefall. The future of fact-based news as an institution is presently uncertain. Hopefully news media 2.0 will emerge as well. In its absence, consumers will need to be wiser, more independent, and resilient in the face of falsehoods, deep fakes, and biased and unreliable reporting.

Characteristics of Democratization

Fig. 2 Democratization (Graph by author.)

> Democratization can be seen at work everywhere once you learn to recognize it.

Organizations and professions that deliver the characteristics described in this section (i.e., making everything available to everyone) will prosper. Knowing how to associate with such organizations, or to create your own, will make or break careers and lives.

Informed, Integrative Choice: Consumer choice is expanding exponentially. Organizations have responded with systems that provide more information, respond to consumer preferences, aid in product search, and provide consumer reviews of their experiences. Examples include Netflix and Amazon Prime for streaming services, Tripadvisor and Expedia for travel, and Carfax for automobile service history. However, these are early versions of future advanced organization-consumer collaborative systems that will be discussed in future chapters.

Customization: Consumers want custom-designed, personalized results instead of standardized commodity goods and services. They want to become what author Alvin Toffler called "a market of one." Examples include customized automobile settings and features, home and auto entertainment, career paths, lifestyles, and a closet full of affordable, bespoke, perfectly fit clothes designed by the consumer themselves or by a designer of their choice.

Direct and Instant: Every possible intermediary step between the consumer's decision and fulfilment will vanish. Most transactions will become virtually instantaneous. Intermediaries will disappear or assume adapted forms and consumers will benefit. For example, the role of commission-based agents is being revolutionized across industries including travel, real estate, and insurance. Typically, commodity services are migrating to the Cloud with little to no agent involvement, while high-end, customized, complex transactions remain agent-based but with constant downward pressure on commissions. Expedia is an example in travel. More people book their own commodity travel, but complex customized services remain available through fewer but stronger specialized travel agencies. Credit cards and online cash transfers increasingly eliminate mail, stamps, and lost time. More and more retail operations will replace middlemen with online shopping and ultimately with 3-D printing for many goods.

Collaborative: Consumers will have a more powerful voice and greater influence in the design of products or services. Increasingly, customer ratings and reviews are one of the most important factors in purchase decisions. Many tech companies already involve customers or extensive customer feedback in product design. Social media influencers are customer-agents who influence purchasing decisions.

Maturational: Democratization is achieved in stages with time and experience, much like personal maturity. Consumers may initially misuse their newfound empowerment. For example, by writing fraudulent or punitive product and service reviews, having egregious social media behavior, and neglecting the opportunity for human intimacy by obsessing over smartphones.

These behaviors will likely mature over time as new social mores emerge.

Adaptability Challenges for Market Leaders

Organizations that remain tied to industrial-age practices will find it far more difficult to adapt to the forces disrupting them than those who are aggressively democratizing their products and services. Compare the old General Motors with Tesla, General Dynamics with SpaceX or Virgin Galactic, and Uber with the taxi service industry it disrupted. There are thousands of other examples across every industry and in all sizes of companies.

Organizations will be less adaptable to the degree they are large, old, centralized, management-dense, diversified, market dominant, and confident in their superiority. Or more colorfully, the rope that an organization climbs to success forms the noose that hangs it unless adaptation to change is swift.

Organizational Disruption Risk Indicators

The unhealthiness of organizations becomes visible when they fail to respond to growing consumer dissatisfaction and increasing competition. Consumers are, after all, an organization's nutrition. If suddenly there was less food available, individuals would react immediately. Yet organizations and institutions often seem to be deaf and blind to their own looming destruction. We've all witnessed this behavior with our employers, competitors, or with institutions. There are eight organizational indicators that predict disruption.

1. Mission loss or confusion occurs when organizations are

run more for the benefit of insiders than customers. It is a chronic symptom in many organizations and usually reflects a failure in governance to hold management teams accountable for mission drift.

2. Sustained stakeholder imbalances occur when an organization fails to maintain the delicate balance between various stakeholders' interests. Each stakeholder is necessary. Each has conflicting demands that must be balanced to sustain a healthy organization. Customers want the best product or service at the lowest price. Shareholders want a return on their investment. Lenders want their debts repaid. Suppliers need a fair price and contractual terms to stay in business. Employees seek a just wage and employment stability. If an organization chronically fails to maintain a balance, it becomes impaired because one or more of the stakeholders will refuse to cooperate or provides tepid support, which will be reflected in failed promises to all the others.

3. Price increases become unsustainable when there is a widening gap between the consumer's ability to pay and the value received. In this scenario, a breakdown in the original consumer value proposition has likely occurred.

4. Competition is essential for healthy organizations. Confident organizations believe in their unique value proposition and recognize that both they and the consumer benefit from competition. The best sports teams, and individual performers, are created by strong competitors. We all do better in our daily performance when we are challenged; otherwise, we slack off. Monopolistic organizations, however, fear competition because they doubt their own value proposition or, like unchallenged people, they get lazy.

5. Lack of competition creates an insensitivity to consumers if they have nowhere else to go. In the end, lack of competition and insensitivity to consumers is always deadly to the organization itself. Every good company welcomes competition as healthy for them, their industry, and the consumer.

6. Denial is a refusal to accept competitive vulnerability, a failing consumer value proposition, or a refusal to take the necessary actions to address them.

7. Polarized thinking creates an "us versus them" or "my view versus yours" attitude in which one's ego displaces good judgement and open-mindedness. It usually occurs when winning becomes more important than succeeding, or when one doubts the virtue of one's own position.

8. Corruption is a telling sign of organizational distress. We all respond with concern when a company we do business with or an institution we rely upon is charged with corruption. Exposure of corruption is an essential role of institutions like media and government, with government having the additional responsibility of using the courts to pursue remedies. Democratization will enable consumers to become more aware of, report, and react collectively to corruption, which will become more important as institutional responses become less effective.

CHAPTER 3

Demographic Change and Neutralizing Distance

> "Aging is not lost youth but a new stage of opportunity and strength." —Betty Friedan
>
> "Technology gives us the facilities that lessen the barriers of time and distance—the telegraph and cable, the telephone, radio, and the rest." —Emily Greene Balch

IN THIS CHAPTER, WE EXPLORE the possible effects on your life and career of an aging society and the simultaneous neutralization of distance. For the first time in human history, birth rates are falling, and life expectancy is extending in most countries. The consequence is a world with a much greater proportion of elderly to young people. In parallel, distance-neutralizing inno-vations are reducing or eliminating time spent getting from one place to another and making the world available to everyone as never before.

Demography

Unprecedented population changes are underway in every city and country and for the world as a whole. Changes in total pop-ulation and in its composition will profoundly affect the careers and lives of everyone in the twenty-first century, as illustrated by the following questions.

If the ratio of older to younger people doubled or tripled, would there be enough younger people to finance retirement and health-care safety nets for the elderly through taxes and insurance pre-miums? Would we have enough younger people in the workforce to keep things running?

If birth rates continue to fall and longevity increases, won't the population decline? Since economic growth is fueled in large part by an expanding population of consumers, how will econ-omies adapt to a declining population? Would companies—and stock market investments in them—suffer from fewer available customers each year? Would home values decline if there were fewer possible buyers? Would government, education, and reli-gion struggle if there were fewer taxpayers, students, and faithful each year?

On a personal level, you might ask yourself: *If I live longer and healthier and can live and work from anywhere (WFA), will I want to retire at sixty-five? Can I afford to? How much more money for retirement and health will I need for every year longer that I live?*

Can I find work if I choose not to retire?

These are not futuristic possibilities but present-day increasing concerns in half the countries in the world. However, demographic change is not like the fast-paced plots of disaster movies. Demographic change occurs over decades, allowing each of us, and whole societies, time to adapt. But to date, we have not adapted collectively or individually, so it is time to begin. Until about twenty years ago, these historic demographic changes went largely unnoticed. Even today, world leaders and the news media rarely discuss their significance. Most people are unaware that they will need to adjust their plans for career, retirement, lifestyle, health, finances, and investments.

Rise and fall of annual population growth

The world's population grew less than 1 percent annually until the First Industrial Revolution in the 1700s. Annual population growth then doubled and peaked in 1968 at 2.1 percent. A decline followed, reaching 1.08 percent by 2019.* The rate of population growth reversed because of a rapid decline in birth rates worldwide. Simultaneously, life expectancy extended worldwide in most countries. In both cases, the developed world led the trend, but the less developed world is following and closing the gap. The COVID-19 pandemic reduced life expectancy, but it was not enough to change the overall trends, and life expectancies are expected to rebound and continue to increase.

---

*   Max Roser, Hannah Ritchie, and Esteban Ortiz-Ospina, "World Population Growth," Our World in Data, 2013, revised 2019, https://ourworldindata.org/world-population-growth.

Increasing life expectancy

Many people assume that medical science alone extends life expectancy. It certainly has performed miracles, but adequate food, clean water, accessible public health systems, modern sanitation, traffic safety, drug and alcohol education and treatment, fewer and smaller wars with less loss of life, and a long-term reduction in violent crime have all played an important role.

> Life expectancy could surge in the future from advances in healthy longevity science and practices and cultural attitudes toward health and longevity.

Some longevity professionals believe that improved treatments and cures for many age-related ailments such as Alzheimer's disease and a wide range of cancers are on the horizon. They are equally encouraged by research that is exploring the aging process itself, rather than relying solely on research into the diseases of aging. Also, more people of all ages in developed countries are becoming health conscious. A cultural norm seems to be emerging that age seventy is the new fifty and age ninety is the new seventy. Products and services that support this new cultural norm are increasing.

Declining fertility rates

Total Fertility Rate (TFR) is the average birth rate per female in a population. A TFR of 2.1 births per female is required for a population to remain flat (i.e., zero growth). According to UN estimates, almost one-half of the world's population lives in countries with a TFR below 2.1. Some heavily populated countries including China, the United States, Russia, and Brazil are at 1.7 TFR. Countries such as Japan, Thailand, and most of Western Europe are experiencing a rate of 1.5. Thirteen countries including Spain, Italy, Singapore, and South Korea are below 1.3, with Singapore and South Korea at 1.0—ostensibly a population freefall. The further the TFR gets from 2.1, the fewer generations until the country's historical natives disappear. It is painful to imagine Italy with few Italians and South Korea with few South Koreans. Yet to date, no country has successfully reversed a declining TFR. Birth rates have steadily fallen in less developed countries as well, but they remain above the population replacement rate. Only Africa, parts of the Middle East, and South Asia are above 2.1 TFR. Africa is the only continent likely to continue population growth after 2160, and that growth will come from momentum of the current population, which some estimates suggest will play out by 2100.

Causes, duration, and reversibility of falling birth rates are cloudy, debatable, and controversial. There are many explanations (some sources suggest over forty) for why fertility rates have fallen so far and so fast. Some of these include birth control, starting families later in life,* better educated professional women with less time for large families, reduced infant mortality, and reductions in agrarian families that rely upon children to work the farms. The bottom line is there are simply fewer reasons for people to have large families, including the cost. Middle-income married couples in the United States can expect to spend on average, including inflation, $284,570 raising a child to age seventeen—*excluding a college education.*†

---

*   Quoctrung Bui and Claire Cain Miller, "Age That Women Have Babies: How a Gap Divides America," *New York Times*, August 4, 2018, https://nyti.ms/33flhNg.

†   Mark Lino, "The Cost of Raising a Child," US Department of Agriculture, February 18, 2020, https://www.usda.gov/media/blog/2017/01/13/cost-raising-child.

In addition to the falling TFR, there are concerns and much debate regarding male fertility. Male fertility has been steadily declining over the last four decades, according to an article published by Oxford University Press in 2017.* The authors reported that there was a decline in sperm concentration and total sperm count in samples collected from 1973 and 2011. Significant differences at the time of the study were limited to North America, Europe, Australia, and New Zealand. More recent studies conclude that sperm counts have fallen over time, but the causes are unclear, as is the relationship to birth rates. Declining sperm counts remain an unresolved concern.

There is also speculation about other causes for declining birth rates. In nature, species sometimes instinctively increase or decrease their offspring to adjust to their environment. Some experts wonder if this natural process could be at work in humans in declining birth rates. Other experts say that such an answer is too convenient or deterministic. Even if there is such a "natural" explanation for falling birth rates, it calls into question whether life expectancy increases in response to falling birth rates or the reverse.

The net effect of declining fertility and extending longevity

Life expectancy increases and falling birth rates will cause population losses and an aging population, but the degree of change by country, region, and globally remains to be seen. Less developed regions of the world will continue to experience population growth for at least three decades while developed regions decline. The ratio of older people to younger people will increase in most of the world, and ultimately worldwide with attendant challenges.

---

*   Hagai Levine et al., "Temporal Trends in Sperm Count: A Systematic Review and Meta-regression Analysis," *Human Reproduction Update* 23, no. 6 (November-December 2017).

However, the population of younger people will be concentrated in Africa, the Middle East, and South Asia. If the world's TFR falls below 2.1 after 2160, world population could decline for the first time since the bubonic plague. However, predicting specifically when and if global population will decline is a mathematical probability exercise that nothing but time and experience will corroborate.

Population estimates are intensely controversial. Projections of population growth, decline, or composition affect powerful economic and political interests. They influence decisions on hotly debated issues such as the environment, immigration, farm subsidies, the age requirement for retirement and healthcare benefit eligibility, pro-life and pro-choice debates, and capital punishment, to name a few. We rarely hear about population decline risk from our institutions, especially government and the media because of their relationship with these hot-button issues and the uncertainty of population projections.

There is a curious and important anomaly surrounding extending longevity that we can all observe. The stages of life (childhood, adolescence, early adulthood, etc.) seem to expand proportionately to extended longevity. Today, many adolescents and young adults seem to take longer to mature. Families are having fewer children later in life. Midlife is extending to sixty, seventy, or eighty, and living to over one hundred years will probably become increasingly common in the not-too-distant future. Each young person today can expect not only a longer life, but longer stages of life as well.

Distance Neutrality

Shrinking distance as a democratizing force will benefit commerce, education, travel, electronic connections, and access to experiences. The innovations that overcome distance as a limitation will be available to everyone, including a vast part of the world's population that was previously excluded from its benefits.

Imagine a world where you can travel anywhere faster, safer, and less expensively than at any time in history. Now, imagine reaching almost anywhere in the world in thirty minutes, like Alisha and her family in Chapter One. New Reality travel will include rockets, hypersonic aircraft, hyperloop tracks, self-drive vehicles, personal and Uber-like taxi drones, and maybe a trip into space.

Distance as a consideration, or challenge, will be neutralized through these and other transportation innovations. Equally important, physical travel will become less necessary as virtual, augmented reality and hologram experiences become available and become adept at making meetings at a distance approximate those in person. Permanent, sizable reductions in the number of commuters on the roads and in the number of workers in commercial office spaces resulted from people working at home during the COVID-19 pandemic. Work from home (WFH) or work from anywhere (WFA) is good for the environment, the cost of doing business, and for commuter health. It was enabled by relatively crude technology and innovative people who cobbled together solutions. Given that these reactionary, unplanned solutions worked pretty darn well, imagine what can happen when ingenuity takes hold to improve work-at-home and service-from-home experiences. Imagine Zoom version 62.7, where the meeting or destination experience is so real that you feel you are present in the room with the other participants, or that you are walking through a Brazilian rain forest or circling one of Jupiter's moons. Visualize Cloud-based shopping in which a hologram of the sofa you are considering is projected into your living room in any color or fabric design you desire.

Consider the possibilities of a thought-based brain-to-computer interface (BCI). Travelers will be able to experience in real-time the senses of an Amazon rain f orest explorer or a Mount Everest climber. A thought-based BCI would eliminate the need to augment your reality. Instead, you would simply share someone else's reality. Your future job may use BCIs to collaborate with coworkers, suppliers, and customers to design new products.

You will likely experience these innovations in the next three decades. Travel, business, education, medicine, and the purchase of goods and services will lead the way. Sooner than we might expect, space exploration will become as common as exploration of the Americas in the sixteenth century, enabled in large part by distance-neutralizing innovations.

With every advance comes choice and new decisions. Each of us will need to choose between virtual and physical in-person experiences. Distance neutrality is upon us and is transforming society at blinding speed. It's very likely you are already making these decisions, and more are on their way soon.

Advances are coming soon—very soon. In November 2020, following 400 unoccupied test runs, Virgin's Hyperloop safely transported two employee passengers in its Pegasus vehicle at high speed along its 500-meter Las Vegas test track. Vehicle speeds exceeding 240 miles per hour have been achieved. Virgin says its commercial system will propel passengers at speeds of over 670 mph using electric propulsion and electromagnetic levitation, under near-vacuum conditions. Virgin says this will become reality in years, not decades.*

In the 1980s, few people envisioned autonomous vehicles. In 2004, DARPA's (Defense Advanced Research Projects Agency) self-driving car competition kick-started the race to create road-worthy autonomous vehicles. Today, there are a handful of autonomous vehicles on Phoenix, Arizona, roads traveling without backup safety drivers. Soon self-driving cars will be operating in major cities throughout the United States and developed countries. Within a decade or two, they will likely represent most vehicles on the road.

Today, non-military drones are used daily for photography, search and rescue operations, and parcel delivery. Air taxis will follow once safe use of low-altitude skyways is perfected. Many companies worldwide are looking to enter this market, some targeting flying cars for individual owners. Some pundits and companies predict that the earliest air taxis may appear by 2023; however, the late 2020s or early 2030s seems more likely.

"Cloud-travel" is moving faster. Entrepreneur and founder of Oculus VR, Palmer Luckey, says: "In the past, before phones and the Internet, all communication was face to face. Now, most of it is digital, via emails and messaging services. If people were to start using virtual reality, it would almost come full circle."

Humans are social creatures; something magical happens when we are in physical proximity. Most people prefer face-to-face interactions, no matter how real the electronic interface may feel. Intimacy, or kinship, unfolds during a physical encounter—an unseen but palpable electricity that cannot be fabricated by any other means…at least not yet.

---

* "World's First Hyperloop Passenger Test," Virgin Hyperloop, November 9, 2020, https://virginhyperloop.com/.

In her book *The Girls' Guide to Hunting and Fishing,* author Melissa Bank perfectly captured the relationship between human connection and location in her statement: "Dante's definition of hell: proximity without intimacy." But the reverse of Dante's definition is the challenge of the future: to perfect intimacy without proximity. Each of us, and the technology companies that make distance less relevant and travel optional through electronic means, will need to achieve that perfection to take full advantage of distance neutrality. The technology companies attempting to make travel optional by optimizing digital communication alternatives, and each person seeking the best possible digital relationships will need to perfect intimacy without proximity.

The COVID-19 pandemic taught us that those virtual transactions were easier and sometimes more effective than many traditional face-to-face transactions. Physician visits, working at home versus the office, business and personal meetings, remote learning, and even therapist visits proved equal or better than the face-to-face experience for many people. Already, experts are studying why electronic communications work for some people and not for others and why some virtual transactions are effective and others are not.

Our personal experiences prove that intimacy without proximity is possible, though it may not always be easy. Many of us have family or friends living at a distance, yet we immediately re-establish intimacy with them on the phone, over FaceTime, or during Zoom calls. In most cases, these were intimate relationships previously. But in times past, people met and became intimate friends; some fell in love, all through snail mail, sight unseen. That practice continues, only today the protagonists use email, text, and dating apps.

The "intimacy without proximity" challenge is likely to be solved in time because fortunes will be made, and significant productivity gains depend on its success. Cloud-travel products will prosper if the industry gets it right. Space travel will become less daunting because travelers will be able to have meaningful connections to the families and friends that they left behind. Business travel will occur only when there is an overriding need for proximity. More importantly, virtual travel and experiences have the potential to be *available to everyone,* including those unwilling or physically or financially unable to travel.

In the previous three chapters we explored the forces of change. We examined how innovation accelerates change and creates resistance, which in turn increases volatility. We reviewed how democratization and innovation seek to make *everything available to everyone* and how the process disrupts or deinstitutionalizes organizations that fail to adapt, adding to volatility. We also saw how demographic changes and distance-neutralizing innovations will expand our available time and make new experiences and opportunities available to everyone, even as innovation and the pace of change use that time to make us busier. In the next chapter, we will explore how these forces converge in our daily lives to create *a New Reality.*

CHAPTER 4

The New Reality—
Better or Worse?

> "To survive, to avert what we have termed future shock, the individual must become infinitely more adaptable and capable than ever before. We must search out totally new ways to anchor ourselves, for all the old roots—religion, nation, community, family, or profession—are now shaking under the hurricane impact of the accelerative thrust." —Alvin Toffler, *Future Shock*

THE POTENTIALIST: YOUR FUTURE IN THE NEW REALITY OF THE NEXT THIRTY YEARS is about adapting and succeeding in a very different world than the one we know today. This chapter identifies ten new world realities that will result from the collision of forces of change with everyday life. At the end of the chapter, we will explore whether you will look back at today in 2050 and feel that things were made better or worse in the intervening three decades.

Looking Ahead to the New Reality

New world reality #1

The world will become your stage in the next three decades, but it will challenge you to understand who you are and where you stand as everything about you becomes transparent.

Who we are and how we live is already available to anyone with access to the Cloud. We live in a virtual fishbowl that will grow to include everyone on the planet in the next few years. Deciding how to present ourselves to the other fish can be an exciting opportunity or a nightmare depending upon the decisions we make.

People on social media today express views and opinions, tell stories, or perform. Effectively they are saying, "I am here. See me. Hear me." Some critics blame self-absorption or narcissism. Certainly, much of today's content is puerile, but everyone is experimenting and learning how to use this new Cloud medium—for the first time in their lives, they can have an enormous audience.

Like never before, forces of change are pushing us to take responsibility for our lives: to think for ourselves; provide for ourselves; and take responsibility for who we are and for who we want to become—and ultimately for reaching our potential. We will have more options, more choices, more opportunities, fewer restrictions, and the ability to get faster and more direct results with fewer intermediaries.

How wisely you choose to present yourself today in a world of unprecedented transparency and visibility may well set the course for your life and career.

New world reality #2

You will be superhuman, but will you make the best of it?

Within three decades you will be superhuman—at least by historical standards. That may be a difficult concept to accept. Most of us do not think of ourselves as powerful, much less superhuman. Some time ago we began integrating humans with automation and there is no going back, nor is there any theoretical limit to the integration that might be achieved. Even with the powers we already have, we would have appeared godlike to our great-grandparents. Imagine if they had witnessed us FaceTiming with a smartphone. Imagine their difficulty in comprehending the Internet's ability to answer almost any question or how GPS empowers us to never get lost.

It is equally challenging for us to imagine being equipped with a voice-based or thought-based brain-to-computer interface (BCI) that will enable us to function at the level of genius, allow us to share experiences with others in real time, or to house our bodies in exoskeletons that enable almost unlimited strength. Yet this reality is likely during your lifetime, depending on your age and how fast the technologies become widely available. However, there will be tough choices and decisions along the way with consequences that Susan's story illustrated in the prologue.

Another "superhuman" trait is that you will live longer, perhaps much longer, than your ancestors. Increased life expectancy has the potential to pose difficult choices, as in Peter's story when he considered life-extending therapy, but also to create expanded opportunity for meaningful lives as you have additional years and decades at the peak of your wisdom, maturity, and experience.

Add to all this the fact that lifelong learning will become personalized, available anywhere and perhaps virtually free or at least far less expensive than today. It will be designed specifically for what you wish to learn and how, where, and when you learn best.

Enhanced intellectual and physical power, a much longer life, and unlimited access to lifelong learning and knowledge have the potential to make you the most powerful creature that has ever lived. The kicker is that you will need to decide how to use that power. Without wisdom you could do more harm than good to yourself and others. Used wisely, over a longer life, you could change the world.

New world reality #3

New forms of communication will enable instantly shared multisensory experiences unrestricted by number of participants or distance.

People and societies have evolved communication since cave dwellers drew crude images on their walls and told stories by firelight. The written word has been a boon to humankind for over 600 years and provided a wealth of accumulated knowledge available to everyone. Reading and writing allowed us time to pause, ponder, and expand our consciousness because we controlled the pace and sources of information. Reading sparked our creativity and growth by engaging us in mental and soulful experiences. Today, fewer people read, even as we see greater numbers of books published.

Electronic communication via the ubiquitous smartphone has expanded and enriched human relationships, but with inevitable downsides (resistance) that accompany all innovation. For the first time in history, we can see as well as hear people on our phones, tablets, and computers unrestricted by geographical distance. Texting, social media, and podcasts encourage brief, condensed communications and "bullet points" instead of paced exposition. With its many options, modern-day communications technology has enriched human relationships, and we have only witnessed the tip of the iceberg. New forms of communication will create shared simultaneous experiences more powerful than anything we have yet experienced. Instead of reading the same book or watching the same movie, an unlimited number of people anywhere in the world will be able to use all their senses to share experiences, collaborate, and deepen relationships. But like all technology and change, you will need to make wise decisions on its use and when to adopt it as it becomes available.

New world reality #4

Relationships will become more important, not less.

The prevailing view today seems to be that digital communications will deteriorate human relationships, but the opposite is more likely. Younger people are products, in part, of the communication methods of their time and have a comfort level with them that older people never manage to attain. Many adults in the 1950s believed that television would destroy relationships. Those adults had conveniently forgotten that their parents said the same about telephones. Most everyone will become accustomed to working and living in the Cloud instead of in person in the next three decades. We will learn to build relationships with people in the new forms of media just as we did in the old. Future communications technology will be designed to encourage relationship-building through methods like shared experiences. However, we will always treasure physical connections no matter how convenient and empowering other forms of communication become, and in-person relationships will become even more valued as we spend more time in the Cloud. The challenge will be to learn how to develop intimate relationships, essential to career and life, when electronic communications are easier, cheaper, and more readily accepted.

New world reality #5

Your work will become more meaningful, enjoyable, and better integrated into your life.

Changes to work over the next three decades will be the most sweeping since the First Industrial Revolution in the 1760s. The changes will affect every person in some way. If you are in your mid-fifties or younger, your work and likely your career and job will be redefined multiple times. If you are over fifty-five years old, you will likely interact with new workers who you may never meet in person and some who will be artificial intelligence, robots, replicants, or androids.

The sweeping changes in work are being driven by the convergence of the change forces discussed in the prior three chapters. Innovation—especially artificial intelligence, robotics, and the Cloud—will eliminate many jobs, careers, and professions that are repetitious, dangerous, or physically demanding. It will, however, create new ones that will be more creative and involve problem-solving and complex human relationships. Shortages of younger workers in developed countries will accelerate the need for automating many jobs and create incentives for older people to work longer. Distance-neutrality innovations will progressively enable most jobs to be performed remotely, reshaping lifestyles and changing industrial-age employer-employee relationships. But even more importantly, technology will power new types of collaborative work, problem-solving, organizations, and products and services that we are only now beginning to envision. I will explore these in the next chapter.

Economies, since the First Industrial Revolution, have competed on labor cost and quality, raw material availability, and transportation to move materials and products efficiently. The Cloud economy, advanced mobility, demographic changes, and innovations like 3-D printing could turn all that on its head and reshape economies and institutions as we know them.

Work matters a lot to most people. It can simply mean earning a living, or it can offer enjoyable camaraderie with coworkers or customers. Some people define themselves by their occupation. A large share of the population, perhaps even the majority, derive meaning from the work they do, including some or all of their life's purpose. Work consumes more time than anything else we do except sleep. Those who love their work tend to love life—and the reverse is also true.

The significance of changes to work becomes clear when you consider all the previous points collectively: work's importance; the sweeping nature of the changes; and their effect on people, your career, employer or business, and community. This alone should be reason enough to understand and prepare in order to thrive in the New Work. Assisting you in that quest is the goal of the following chapters.

New world reality #6

Clear priorities will matter more than ever.

The New Reality will cause you to reevaluate your priorities to make the best use of time. Time issues are always priority problems in disguise. Innovation speeds up life and makes it busier, even as it increases available time elsewhere, such as through distance neutrality and extended longevity. An example that we can all appreciate is the smartphone. It is a truly miraculous device that changed life worldwide and, on the whole, made it better. But for all its miracles, it made life busier, not less. Innovations create new choices, more decisions, expand our options—all of which take more of our time and attention, not less. Even when innovations save time, we usually fill it with more activity rather than relaxing or sleeping more. Of course, we all make these trade-offs, but we make most of them unconsciously without a clear, deliberate set of priorities. Having a conscious, clear understanding of what you value most has never been more important.

New world reality #7

Lifestyle will integrate family, work, and dreams as never before.

The choices for where and how we live and what work we do will grow exponentially over the next three decades. Start asking yourself now what lifestyle you would choose for yourself if there were few if any limits.

In the past, most people lived their entire lives in their birth country near their childhood home, close to where they worked, in the best home they could afford, and near good schools. They faced several constraints that held them in place, including their work, social status, education, money, family ties, limited travel opportunities, access to innovation, and a raft of other cultural barriers.

Those constraints are disappearing. Mobility options are neutralizing distance. More jobs and careers can be carried out from virtually anywhere in the world and more will become "work from anywhere" (WFA) as time passes. Language and cultural barriers are being lowered by automation and connectivity. Future virtual travel experiences will expand cultural experiences to an exponentially larger population for whom those experiences would have previously been impossible.

The COVID-19 pandemic offered a glimpse of this new world. Work from home (WFH), WFA, and distance learning were tested using only makeshift technology and practices. In-migration increased from cities to rural areas and between states. The future will likely see people changing countries as readily as they change states or provinces today within a country.

Adopting a lifestyle without limits will require some tough decisions. Choices will include financial flexibility over asset acquisition, experiences over possessions, mobility over stability, and renting versus owning. Many people will choose to live traditionally, but there will be compelling new choices as well.

New world reality #8

Life and career in the Cloud will redefine life as we know it.

If you currently spend 20 percent of your time in the Cloud, your future is probably 80 percent or more. We will live, work, shop, barter, travel, and relate to each other in a digital world made possible by the technologies discussed in earlier chapters. There is no historical analogy to the Cloud, Internet of Things (IoT), or metaverse. It's a new world, a New Reality, the extent of which we are only just beginning to fathom. The world economy will become far less physical and more digital and digitally dependent, and that includes money. Already, people carry far less cash than in the past. Spurred on by the pandemic, more of us bank, deposit checks, and move money using our phones. Digital currencies are finding a place in all this, as discussed in future chapters.

Barring unforeseen circumstances, you will soon be Cloud-dependent for almost all essential functions. Currently, a few hours without electricity causes havoc in our careers, lives, and communities. Imagine becoming Cloud-dependent and access to the Cloud failing for any number of reasons, including electrical failure. We will be paralyzed without the Cloud, in a similar way to how our body can be paralyzed by a stroke because the Cloud acts like a neural network for the planet and every person on it. Wild, huh?

Many issues need to be addressed, and hopefully will be, before society becomes totally Cloud-dependent. Vast improvements in privacy and security are needed, including protection against cyberattacks by hostile states and malcontents in basements. Stable, independent Cloud governance will be crucial but challenging to devise. A comprehensive risk management and backup system will be devilishly complex. The Cloud offers such huge promise that the financial investment to keep it secure should be a no-brainer, but we cannot count on that happening. Nothing in history has had anywhere near the potential to create a flatter, less siloed, more connected world economy and society. The challenge for you will be deciding, to the degree possible, how fast and to what extent you become Cloud dependent and what you can do to protect yourself and your career, or employer, from the risks.

New world reality #9

Reacting and trial and error won't work so well anymore.

We spend more time reacting, learning, unlearning, and re-learning than in any other form of growth and development. That was fine when life was slower, but no longer. Accelerating change reduces intervals between innovations, resistance, and turbulent events. That compresses the time we have available to react, reassess our actions, and learn from our mistakes. Our choices in everything are also expanding. You will need to make many more decisions, that are more complex, far faster than previous generations. You will be hard-pressed to continue trial and error as in the past.

We human beings are a curious, hilarious bunch in many ways, especially in how we dislike thinking ahead and prefer trial and error and reacting. If aliens are watching us, they must be laughing. For example, instructions come with everything, but do we look at them? Oh, heck no! Or, if we have a major decision to make and have experienced people eager to assist, do we ask for their advice? Not on your life. If our GPS fails to work, do we stop and ask directions? Not if we are male. I'm convinced that humans developed flat foreheads from millions of years of slapping their palms to their heads over last-minute reactions and trial and error mistakes.

We need not abandon our old ways entirely. But we need to act much more thoughtfully and deliberately when adaptation cannot wait until the last minute and the consequences are severe. The remaining chapters in this book were developed for just that reason: to help you get ahead of the game where it matters. If "react and solve by trial and error" is at one end of a decision-making spectrum, and "think before you leap" is at the other, we should be aiming for the sweet spot in the middle where we don't overthink things, but also avoid unnecessary risks.

New world reality #10

Sweeping societal changes will alter the fabric of daily life and belief systems.

Today's deeply ingrained attitudes toward such things as aging, nationality, gender, race, personal lifestyles, and education will fade or seem strange to observers in 2050.

Retirement by choice for some people, and necessity for others, will likely be shortened to a few years at the very end of our lives. Established, even revered, vocations and professions will disappear. Venerable organizations and institutions will reinvent themselves or become irrelevant and cease to exist. Entirely new lifestyles and urban environments are likely to appear as office and retail spaces, manufacturing, and commuting are reinvented. Fewer people will live in the country in which they were born and where they have citizenship. People may become increasingly spiritual though less aligned with any specific religion. Some of the best educated, wisest people on earth may never have set foot in a physical classroom. News may come to rely on independent, unbiased journalists instead of corporate media outlets. Far more is unpredictable than predictable but expect a major evolution from today.

Will the New Reality Be Better or Worse?

Having read to this point, do you think things will be better or worse, or a little of both in the next three decades? If you feel optimistic, then you are in a minority, even if you temper your optimism and accept that there will be tough times ahead caused by unintended consequences, turbulence, and Black Swan events that seem to come out of nowhere.

> Pessimism about the future prevails among people in developed countries, even as they live better than people have ever lived.

It is hard to identify why so many are negative about the future, especially given the overwhelming evidence to support that life in general is getting better. History shows that the arc of civilization has been upward for over three million years. For instance, living conditions have improved enormously since the First Industrial Revolution. It would be hard, even among pessimists, to deny that people enjoy a better standard of life than at any other time in history. Unconvinced? A few hundred years ago brutal executions were public celebrations with children present. Animal cruelty such as bearbaiting and cockfighting were popular local "sports." Few roads in the world were safe to travel without some sort of security. Wars were a way of life, regularly almost wiping out entire generations.

Society tolerates none of this today and cannot fathom doing so. Humankind worldwide is measurably better off and more civilized today. We are on the precipice of a giant leap forward that could solve some of humankind's greatest historical challenges—in your lifetime! After three million years of civilization progressing inexorably forward, why would it falter now? Probability alone would suggest a fifty-fifty chance of a better future, but the long arc upward and the recent dramatic acceleration surely push the odds well into positive probability range.

Contemporary authors and researchers have been baffled by the negativity. Their literature describes how human nature tends to overlook the positives and overstate the negatives. *Factfulness* by Hans Rosling makes the case that we distort compelling optimistic data through a lens of negativity. *Abundance* and subsequent works by Peter Diamandis and Steven Kotler, point out that we are on the verge of dozens of life-improving innovations that could elevate the human species. *It's Better Than It Looks* by Gregg Easterbrook provides evidence of an improving world and lays negative bias at the door of a fear-reinforcing environment that I discuss shortly.

Clearly, something causes many of us to become irrationally negative about the future. Do these feelings tell us anything about what the future will really be, or tell us only about ourselves?

Should we care if people choose to be optimistic, pessimistic, or irrational? Personally, I think how people feel about the future is a serious matter because their negativity plays on other people's fear and spreads. We will all need to be balanced and clear-eyed instead of negative, anxious, and fearful if we are to clearly see the opportunities of the future through the turbulence of the times.

Every past industrial revolution has had its doubters and resisters to change. In the First Industrial Revolution they were called *luddites*, a name occasionally still used. Unfortunately, many people with strongly negative views are resistant to facts or arguments that could change their minds. The only way to address irrational negative bias is for the individual to recognize that they are suffering from a fear response to the unknown. Balanced, critical thinking is healthy and productive, but fear-based negativity is a personal and collective enemy. So is blind optimism, but it is so rare that it is inconsequential.

Perhaps there are other deeply rooted causes that explain fear and loss of hope. A growing percentage of the population is older, and older people are traditionally less optimistic. As someone enjoying later life, I believe younger generations deserve better than knee-jerk negativity from those of us they rely upon for wisdom. As we elders become a larger share of the population, we need to shine more light and less darkness into their world. Interestingly, those of us who enjoy frequent interaction with young people are generally more optimistic. When you observe the quality of young people's minds and their enthusiasm on a regular basis, it is hard to be negative.

> We humans are hard-wired to be anxious and fearful—though some of us more than others.

Our long history as hunter-gatherers until 10,000 years ago at constant risk of being prey to animals or other humans made us a fearful, cautious bunch. Our ego and shadow are built for defense; they have not evolved to the twenty-first century's more modest threats. They sense danger where it does not exist while often missing today's real dangers. We tend to distrust the optimist and embrace the naysayer and alarmist. We love horror and disaster movies, but quickly tire of heartwarming stories of love and happiness. Optimism, even if cautious, seems to enrage pessimists and those with dystopian views, perhaps because they fear it might make them drop their guard and become vulnerable.

> Unevolved, fear-based responses to the unknown have helped create a real twenty-first century danger in a culture of manipulation. Fear sells.

Politicians warn that they need our money and votes, or disaster will follow. Special-interest groups catastrophize even the smallest setbacks as the end of civilization. Some religions insist that heaven will be open only to those who subscribe to their specific dogma and organization. Educational institutions warn that our children will fail in life and have no social status without their institution's diploma. News media patiently explains to us (because they believe us to be uniformly ignorant) that without their translations of events, we cannot possibly understand the horrible dangers around us. Social media measures us as losers if we don't have gazillions of followers and "likes," while behind the scenes using algorithmic messaging to exaggerate fear, polarity, and discord. Entertainment media serves us an ever-increasing diet of gruesomeness and coarseness, sprinkled with just a little that is inspiring and hopeful. Many advertisers peddle an urgent need for their products by playing on fear, inadequacy, or imagined consequences of life without it. It takes a sturdy, self-confident person to remain positive in this environment.

Fear and negativity merchants exist because we embrace their messages.

> The toxic cloud of negativity will lift only when large numbers of us reject irrational, fear-based negativity and bias.

How about we turn them off? Stop buying what they are selling. Demand better. Write reviews. Refuse to be polarized. Recognize that much of the division between us today has been manufactured by propaganda that we swallowed. Seek purveyors of wisdom and people we can learn from instead of advocates who simply reinforce our own bias or point of view. Search for and then defend the virtue of different points of view than our own.

Above all, face down negativity by broadening perspective. Our slot on this earth is a tiny slice of infinity. As author Daniel Goleman says, "No one has the perspective to be a pessimist." Each of us has a choice; we can shine light or darkness into the world. The case for light has never been better.

Skills and Mindset for New Reality Careers

THE PROLOGUE OFFERED A GLIMPSE into lives and careers over the next thirty years. Part I, chapters 1 to 3, summarized the change forces that will shape those three decades. Chapter 4 condensed the effect of those forces on careers and life into a ten-point summary of the New Reality.

Part II proposes mindset change and skill development in four foundations for twenty-first century careers: "The New Work and the Collaborative Revolution," "Differentiating Yourself in a

Democratized World," "New Reality Relationships," and "Thinking Like an Entrepreneur."

These four foundations are essential to twenty-first-century careers and are often neglected. They can be improved relatively easily with a shift in mindset and the adoption of a simple methodology that will serve you well for a lifetime.

CHAPTER 5

The New Work and the Collaborative Revolution

> "If you look at history, innovation doesn't come just from giving people incentives; it comes from creating environments where their ideas can connect." —Steven Johnson

THIS IS THE FIRST OF several chapters that illustrate how the "New Realities" and "New Work" will reshape careers and lives by honing certain skills. The New Work, as previously noted, will be more creative, innovative, and involve complex problem-solving and the complexities of human relationships. This chapter addresses problem-solving and innovation through collaboration. A later chapter addresses relationships. As you read the next few pages on New Work and the potential transformative effect of technology-enabled collaboration, imagine how your career and life could change as a result.

The New Work

Your career model will be quite different from that of your parents' and grandparents' and will diverge further in coming decades. Their careers were straightforward: high school or college, jobs with one to four employers, compartmentalized work and homelife, stable social institutions, and typically a short retirement before death (often from neglecting their health). All that is over—the career and lifestyle of your forebears is not going to be your path.

New Reality careers will be defined by the following.

Work will be more meaningful because it will involve creativity/innovation and complex problem-solving and human relationships. There will be more entrepreneurial choices posing less risk, resulting in more people choosing entrepreneurial careers over traditional employment. Most people will find that these changes lead to more rewarding careers.

There will be more choices in how we work, including being either a full- or part-time employee, being an independent contractor, a gig worker, or an entrepreneurial owner.

The Cloud as the workplace, and center of life activity, will enable work from home (WFH), work from anywhere (WFA), remote distance learning, and reduced business travel. This in turn will enable a better-integrated work and home life and reduce work-life balance issues. Even more important, new, enjoyable, and fulfilling lifestyles will become available because our work no longer dictates where and how we live.

It will be necessary to work until later in life to finance a longer life expectancy and to compensate for a deteriorating public safety net. Also, more people will choose to work longer because their work is more meaningful and integrated with their chosen lifestyle. The world will reap new benefits from people working and contributing longer at the peak of their wisdom, judgement, and experience.

You will have more jobs, careers, employers, and entrepreneurial gigs as employers, jobs, and careers come and go faster over a longer work life.

Advanced communication and collaboration tools will facilitate shared experiences that bring people together to solve problems, innovate, form relationships, and integrate work and life as never before (as illustrated previously in Alisha, Susan, Peter, Greg, and Linda's stories). This is perhaps the most exciting of all the potential innovations of New Work and the New Reality and is the subject of the rest of this chapter.

The pacing factors

> Work is redefined today by innovation, especially automation, which will accelerate in the next three decades.

The type of work that automation can assume will expand, and integration of humans with automation will expand the work that humans can do and the problems that they can solve. The pacing factors for change include the adaptability of humans to expanded powers and the limitations of automation itself, such as computing power, storage capacity, communication transfer rates, and network availability. The cost/benefit of replacing human work with automation is another pacing factor. Cost/benefit further defines what specific jobs and functions automation will assume at a given point in time. The easiest work to cost-justify is repetitious, dangerous, and labor-intensive. The most difficult to automate, and therefore cost-justify, are creative/innovative, complex problem-solving and human relationships. For that reason, work featuring difficult-to-automate characteristics is the most likely safe harbor from being replaced by automation.

The unknown in this simple logic, however, is how clear the line between automation-empowered humans and sentient computers will be in the future. This may seem to be a distant scenario, but according to the futurists, if you are in your mid-fifties or younger, you may live to witness the blurring of the lines between humans and automated forms of humans such as replicants or androids.

Workforce transition

Joblessness and resistance to change plagued previous industrial revolutions. It could happen this time as well, but there is a chance that this transition will be less chaotic. Labor shortages are likely in developed countries because of falling birth rates. Developed countries are also where job losses from automation will occur first. The two may offset each other to some degree. If labor shortages are severe, the employer-employee power balance will shift in favor of both younger and older employees. Young people will be increasingly valuable because of their energy and comfort working with automation. Older workers will be in demand to fill essential jobs because there will not be enough young workers to fill available positions. Also, the experiences and wisdom of older workers will be a good fit for demands of the New Work.

We can see examples of older and younger workers simultaneously becoming more valuable in current job trends with airline pilots, accountants, and physicians. Mandatory retirement ages are being extended, or eliminated, to discourage older workers from retiring. Simultaneously, younger people are being fast-tracked to replace retiring older workers. The timeline for job automation is difficult to predict because no one knows how long it will take to overcome the pacing factors of human adaptability and limitations of automation noted previously. Also, the factors that would affect cost/benefit calculations such as labor automation costs are constantly changing. The more gradual the pace of job automation over the next thirty years, the smoother the transition. But of course, the more gradual the pace, the longer it will be before the New Reality benefits can be enjoyed.

Some people are legitimately concerned that workers performing industrial-age jobs that will be eliminated do not have the creativity or education for New Work. Time will tell, but as Gary's case study illustrated, everyone has a unique creative core and passion. And though education is a realistic concern, a large percentage of working adults are well-educated in developed countries and advances in personalized, remote learning will expand educational and job training opportunities as never before. There will always be people who enjoy working with their hands as well as their minds. Traditional blue-collar jobs increasingly use technology and, in the process, are becoming "light blue." These new jobs are as financially attractive as many white-collar jobs. They are also fertile ground for "light blue" entrepreneurs. Examples include construction, home-related service businesses, and automobile maintenance as vehicles become computers on wheels.

Technology-Enabled Collaboration

Mass production of problem-solving and innovation

Most problem-solving and innovation today is carried out by individual companies and independent academic research organizations using relatively small teams, and in some cases by a single individual. New communication technologies and brain-to-computer interfaces (BCIs) can enable innovation and problem-solving en masse—millions of people simultaneously working on a single problem or innovation, or small teams simultaneously working on a million smaller problems or innovations. Think of this development as the potential for mass production of problem-solving and innovation.

Mass-production of goods was the great social benefit of the first revolution; mass production of problem-solving and innovation can be the great social benefit of the Fourth Industrial Revolution. To deliver on this promise, however, people and organizations must learn to collaborate far more effectively than they have in the past. People who previously performed narrow, specialized jobs in relative isolation will need to think and act very differently if they want to excel in collaborative problem-solving and innovation. They will need to think outside the box and collaborate if they want to be stimulated and challenged to create better results. Teams are highly effective at complex tasks; the challenge has always been assembling the teams and getting them to work together effectively.

Automation will make the transition to collaborative New Work possible and easier. Cloud-based technologies such as virtual reality, augmented reality, and holography will simplify the logistical problems of assembling teams in a shared experience. Gaming technology adapted for business use, team and individual automated assistants, and voice and thought brain-to-computer interfaces (BCIs) will help organize the problem-solving and creative work in a way that best utilizes each team member's skills and optimizes their interactions. These technologies can be to the mass production of problem-solving and innovation what steam power was to the mass production of goods.

Many young people are already, unknowingly, developing skills with collaboration technology. You may have dismissed electronic gaming as child's play. I did, too, until I read the book *Ender's Game* by Orson Scott Card. The book takes place in the future. Preteens and teenagers are recruited by the military to practice becoming warriors by playing "games" in which they destroy enemy forces and sometimes entire civilizations in space. The gamers eventually discover that the military lied to them and that instead of participating in a game, they were killing millions of other intelligent beings. Ender becomes the children's leader, and they turn on their military leaders and take over the world with the same skills that their leaders had misused.

Team-gamers are on the leading edge of technology-enabled collaboration through shared experience. Gamers play together, improve their skills, and often form lifelong friendships without meeting face to face. Esports, the new name for electronic gaming, uses leading-edge collaborative technology and is predicted to be a \$1.5 billion industry by 2023. Investment capital is flowing into gaming technology companies as investors recognize the future growth and potential of commercial applications. Many gaming pioneers are now adult employees, employers, and entrepreneurs. They understand what goal-directed collaboration means. They will adapt to technology-enabled collaborative commerce faster than many of their colleagues.

Insights from thought leaders

Highly respected futurists and technologists like Peter Diamandis and Ray Kurzweil and inventor-entrepreneurs like Elon Musk see this future clearly. Diamandis, in an April 11, 2021, blog post, heralded what he called "the age of mass genius." He noted that "Ground-breaking progress in BCIs is driving us closer to Ray Kurzweil's prediction that our brains will connect seamlessly to the Cloud by 2035." Diamandis also offers a staggering insight: "BCIs will amplify both average human intelligence and our access to an instantaneous wealth of knowledge."*

---

*   Elon Musk's Neuralink aims to connect its brain chip to a smartphone app, unlocking a wireless connection to the Cloud. Monkeys implanted with the device can already play video games. Musk aims to implant the chips in humans by the end of 2022. The implant procedure would be similar to LASIK in time and recovery, according to Musk. AlterEgo, created by researchers at the MIT Media Lab, allows humans to communicate with their computers without lifting a finger. The non-invasive device can decode brain signals for vocalization with 92 percent accuracy. The device is currently being tested and may soon allow users to Google queries without lifting a finger.

You as a genius problem-solver and innovator

Notable historical collaborative successes include the World War II Manhattan Project and NASA putting a man on the moon in the 1960s. Imagine being able to perform at this genius level. Then imagine merging your newfound genius with other newfound geniuses into what amounts to a shared collaborative problem-solving, idea-creating mind that routinely achieves what brainstorming and collaborative teams rarely produced in the past. The historically complex and costly assembling of collaborative genius could be just another day at the office for you.

Collaborative products and services

This new technology-enabled collaboration is not limited to how teams work to problem-solve and innovate. It also initiates a new age of collaborative products and services that replace traditional vendor-customer relationships. For example, Uber and other rideshare services succeeded in part because they act as a "convener" in a collaborative process. A willing driver and passenger were matched. Both have information on each other beforehand and the opportunity to rate each other after the interaction. Their communications are instant, using their phones, and it is cashless.

The collaboration involved was revolutionary when compared to the "take it or leave it" taxi service experiences previously available. The same collaborative, democratized intent can be seen in streaming services. Streaming service software collaborates with subscribers to find something entertaining for that specific individual, leveraging past preferences and an enormous film selection. This is in stark contrast to the "take it or leave it" traditional movie theater selections, TV, or cable fixed offerings.

There are countless other opportunities that could disrupt traditional vendor-customer offerings with collaborative alternatives. Two case studies at the end of this chapter illustrate companies founded on the principle of collaborative products and services. These types of opportunities are worth factoring into your career plans and investment portfolio.

Your challenge—don't miss this train

If Diamandis, Kurzweil, Musk, and other thought leaders are correct about genius-level collaboration, it will reshape your life and career profoundly and quickly. You will want to be part of the revolution rather than a spectator.

> A worthy goal is to become known for your collaborative knowledge, skills, and leadership acumen.

To do this will mean staying on top of developments in collaborative technology and management organization and leadership, addressing your own skills, and gaining experience with leading-edge companies, products, and services.

The collaborative mindset

A good place to begin is understanding why everyone struggles to some degree with teamwork or collaboration. There is a fundamental tension between individual and collective interests that affects the motivation to collaborate. The struggle has been the subject of innumerable works of philosophy, political ideology, management theory, and psychological self-actualization.

When asked to collaborate you must reconcile your best interests with the interests of others, an organization, or society. It is possible, but not always easy, to balance the interests. In the past, it has typically taken more effort to collaborate or to create a collaborative organization than to be individually creative or serve as a cog in an industrial organization.

People cooperate willingly and spontaneously when there is collective urgency or when there is the opportunity for the exhilaration of collective success. Most people who have participated in team sports find their teamwork experience created a lifetime of rich memories and relationships. The same is true for family experiences and projects. Collective heroism and sacrifice during the Battle of Britain, Pearl Harbor in 1941, or 9/11 are among humanity's finest hours. We are usually amazed at what we can accomplish together using only crude makeshift solutions and technology. We learn a lot about ourselves, our collaborators, and the organizations with whom we affiliate. Something powerful is unleashed by spontaneous individualism in support of collective well-being.

As an example, every nation has something equivalent to the US Medal of Honor to celebrate individualism in support of collective well-being in wartime. Collective urgency drove the Manhattan Project to prevent Nazi Germany from being the first to possess an atomic bomb. Similar collective urgency scenarios include when a group of people spontaneously rescue someone following an auto accident or come together to fight a fire or respond to a natural disaster. The desire for the exhilaration of collective success was the emotional fuel behind putting the first astronaut on the moon.

> Technology will alter the natural tension between individual and collective action and give the edge to collaboration.

Instead of urging people to cooperate (often against self-interest), technology will systematically remove barriers to collaboration and better align individual incentives. In the past, technology has influenced the balance. Smartphones increased collaborative activity with virtually no resistance.

Team-gaming technology is designed to promote and reward collaboration. The Cloud, metaverse, AI, robotics, and sensors will form the collaborative platform. Human-to-human collaboration will be enhanced by products like Zoom or Microsoft Teams, AR, VR, holograms, and gaming technology. Think of these tools today as version one, and then imagine what future versions might be capable of doing to enhance collaborative effort. Voice brain-to-computer interfaces (BCIs) will enhance collaboration between individuals and teams through specialized automated assistants that coordinate effort and serve as project managers. Finally, thought BCIs will enable the ultimate collaboration through instantaneous shared experiences. The technology is only part of the change ahead. It is an enabler of a mindset that recognizes the inherent productivity and creativity multiplier effect of spontaneous individualism in support of collective well-being. Utilize every opportunity to experiment with the effect and the technology in your work and personal life until you sense what you are after in this type of collaboration.

Learning the technology

A stream of increasingly sophisticated collaborative technology for business will become available in the coming years. This will include advanced versions of applications like Zoom and Microsoft Teams, virtual and augmented reality, 3-D printing, holograms, and gaming applied for use in business. Spend time getting up to speed and then staying abreast of developments in the field of collaborative work as it moves from gaming and entertainment to commercial applications. However, do not stop there; it will be important to learn to use the early versions of these applications before the advanced capabilities offered by innovations like brain-to-computer interfaces (BCIs) become available. If you have not personally experienced virtual reality, experiences are available in most cities. It will be an eye-opener that will expand your perspective of what is coming. If you have a friend or relative who participates in team gaming, especially those using virtual reality headsets, you will be able to gain insight into what collaborative work will become, probably sooner than you ever imagined. What you are trying to avoid is your job and career suddenly demanding a skill that you know nothing about. It is fun to experience and learn about these technologies in any event.

Developing your collaborative personality attributes

Organizations (and communities—and nations for that matter) have always relied upon people who knew how to get things done through collaboration. Those people provide good examples of the desirable qualities a collaborative professional, or leader, should exhibit. These qualities, as listed in this section, were garnered from my observations of highly successful, collaborative professionals and leaders and from discussions with other corporate leaders about collaborative qualities.

> *Seasoned and Resourceful*: Collaborative professionals are the go-to people who get called to address challenging opportunities or troubleshoot when things hit the fan. Their knowledge of an organization, community, or nation—and its

culture, people, and processes—is both broad and deep. They are apolitical, focus on the endgame, and avoid distractions. They are motivated more by the trust placed in them than other forms of recognition.

Thoughtfully Aggressive: Collaborative professionals think like entrepreneurs. They view problems through the lens of opportunity but temper their optimism. They challenge ideas, especially their own, with critical analysis and risk assessment. However, in the end, they are biased toward action and play the odds.

Demonstrated Collaborative Skills: Collaborative professionals lead by persuasion, example, and motivation, even when their authority is unclear or nonexistent, as when leading a group of volunteers. However, they can be tough when necessary, and people avoid disappointing them (sort of like your mom). They are known as creative contributors and collaborators instead of "unquestioning soldiers." They demonstrate their collaborative skills beyond their job and organization, often being asked to solve problems spanning multiple organizations, professions, and communities.

Solid: Collaborative professionals have a relatable personality that inspires cooperation and collaboration. They emphasize common purpose as well as individual contributions, rather than giving orders or displaying authority. People enjoy working with or for them. They are confident without the need to prove anything to anyone, and they inspire others to be confident in themselves. They are calm under fire and manage their own stress and help others to manage theirs.

Inquisitive: Collaborative professionals are known for asking penetrating questions. They like to get to the bottom of things and distrust generalities and assumptions. They are quick studies who thrive outside their comfort zone. They are usually early adopters of technology and master it quickly, then train or encourage others.

How to find leading-edge adaptive and collaborative organizations

What does leading-edge mean? Organizations that master collaboration will reflect it in their products and services, business practices, culture, employees, suppliers, and customers. Leading-edge collaborative organizations will understand the principles in this book of accelerating change and the need to adapt rapidly. Many have been working to become more agile and adaptable for years and will be well-positioned to take further steps as times dictate and as enabling collaborative technology allows. They have been emphasizing collaboration and teamwork but without the supporting technology. Every organization will adapt to its unique circumstances. Fortunately, becoming more agile as an organization and mastering collaboration is more a shift in emphasis for many rather than a major directional change.

They are not so difficult to find. Information is widely available in seminars and business news feeds, in entrepreneurial journals, and in the publicly available information on stock exchange-traded companies. Two of the larger, better-known collaborative companies include Tesla and SpaceX. They seem to do more with less through close collaboration between committed teams. Toyota also has some units within the giant organization that seem to repeatedly set new collaboration milestones. However, it would be a mistake to ignore the early-stage and midsized companies where collaborative innovation is fertile. Start with industries where you have a particular interest. Your friends and contacts are excellent sources now that you have a better idea of the prize. Another way to evaluate organizations is how well they address challenges to adaptability and collaboration.

Organizational Adaptability. Most companies today operate using an organizational model little-changed since the First Industrial Revolution of 1760. At that time, the first large manufacturing companies were being founded, but only the church and the military had structures in place to manage large organizations. The military model was easier to copy, and ex-military officers were available to lead the new manufacturing companies. For this reason, military command and control became the de facto model for running large businesses. Today's organizations unconsciously retain military remnants in their multiple management layers and highly specialized, functional silos and jobs. Command-and-control is ideal for producing uniform goods at high volume. It is antithetical to creative collaboration because it promotes internal competition and misaligns incentives. Interestingly, the military has become much more agile using special forces units built on high levels of individual skill blended into a committed team, while businesses generally have not yet discovered a replicable collaborative model. Organizations that are changing the industrial-age model are better positioned for the future.

These can be recognized by how Cloud-based and virtual their operations have become and the clarity of their mission and value proposition. They are also ambivalent to the location and the relationship of those doing the work (employees, independent contractors, gig workers, and suppliers), and they have positive, collaborative relationships with their suppliers.

Remember that leading-edge, adaptive organizations are less likely to be current market share leaders. As discussed in Chapter 3, size, age, centralization, management density, diversification, overconfidence, and a lack of innovation are often unfortunate side effects of market dominance.

Integration of Specialization. Specialization is essential to mass production of goods and services. It standardizes labor and processes and makes them progressively more efficient. An unintended consequence of specialization is that as organizations become massive, it is difficult to understand the organization in total, let alone integrate its products and services or remain highly responsive. This problem has long been recognized. Management systems, giant customer service call centers, and the dreaded phone tree were invented to make organizations function as a whole. As every consumer knows, they are largely unsuccessful. Most of us would prefer a dental root canal to being trapped inside a phone tree. Customer service issues, product recalls, and data security breaches often result from a lack of integration.

In healthcare, a lack of integration can be deadly. The care for patients with multiple chronic health conditions, most of them elderly, is typically provided by a half dozen or more specialty physicians such as a gastroenterologists, cardiologists, and rheumatologists. Only a small percentage of patient care today is integrated under a single care plan by a supervising physician. Care coordination and integration instead falls on undertrained and overwhelmed patients and families. For all the benefits of specialization in medicine, the lack of coordination and integration of care is a growing and dangerous problem. Technology-enabled collaboration is likely the answer for both corporate and professional integration of specialized functions.

Meeting Changing Customer Expectations. Democratized consumer expectations baffle many companies today. I recently heard a career retailer describe today's consumers as informed, decisive, demanding, and revengeful. I could not understand what he meant by revengeful and asked for clarification. He said that today's consumers will go out of their way to brutalize you in a bad review or gleefully tell their friends how awful you are. I understood what he meant, and I'll bet you do as well.

Futurist Alvin Toffler identified that innovation accelerates change; we get busier and more intolerant of our time being wasted. Consumer expectations quickly go viral and become cultural. We expect choice but it must integrate into what we already have. We expect abundant information to help us make informed choices, but the information must be simple and available on our phones. We expect customized products determined by our actions with little or no personal effort on our part. Transactions must be automated and instantaneous. Free shipping and no-questions-asked returns are mandatory; there must be no penalty for changing our minds.

An organization's reputation for customer service and utilizing customer input is a good indicator of how seriously they are taking democratized consumers. Reengineering products and services with technology-enabled collaboration will likely be the best way for organizations to meet democratized consumer demands.

Eliminating Middlemen and Interim Processes. Another good indicator that a company is forward-looking is whether they proactively eliminate intermediary processes and the number of people between the customer and product or service. As you evaluate affiliating with business-to-business organizations, remember that companies who play an intermediary role between the consumer and the product or service will be under increasing pressure to improve efficiency as companies shorten supply chains and eliminate middlemen.

Rethinking your career game plan

There is a strong argument for rethinking your career given how significantly automation, New Work, and collaboration will change careers and lives. Rethinking your career is worth considering, whether you are early in your career, in mid-career, or considering a late-career restart.

Career planning in the past typically focused on getting a job with a good company and then seeing where it led. Priorities were pay, benefits, company reputation, work location and environment, job functions, growth opportunity, bosses and coworkers, and commuting. All these factors matter, but in the future more care must be taken to avoid career dead-ends as illustrated by Susan's story in the prologue.

The career you should plan is not among those you see today, but what the industrial and commercial landscape is becoming.

> You could well become a Fourth Industrial Revolution and technology-enhanced collaborative genius.

Your career could be in the collaborative mass production of problem-solving and innovation. You could work with automation and connectivity tools over a long career* to solve heretofore unsolvable problems.† Your career and lifestyle will likely be more integrated. Given that scenario, consider the following approach:

1. Consider your innate talents and passions and write them down. Now imagine if they were the focus of your career; how would they enhance the meaning of your life's wor??

2. With your previous answer in mind, look for a world-altering problem or a much-needed innovation that aligns with your ideal career.

3. Determine the role that best fits your chosen life's work (e.g., employee, independent contractor, gig worker, entrepreneur, or key employee at an entrepreneurial enterprise). Take into

---

* These include Zoom, V10 social media, virtual and augmented reality, and holograms, in addition to unannounced technologies.

† Ending subsistence living, worldwide free education, energy-free conversion of saltwater, environmentally neutral energy, freshwater, ending parasitic diseases like malaria, healthy longevity, space travel, connectivity and distance-neutralizing innovations, advanced mobility, robotics, etc.

consideration that the role could change with experience and circumstances.

4. If you are already a professional such as an attorney, accountant, physician, etc., make the decision to remain in that role or change to one that aligns with your decisions in points one to three.

5. Design the ideal lifestyle that aligns with the decisions in one through four.

6. Research the companies and organizations that are devoted to the problem or innovation you listed in number two, as reflected in their mission and value proposition. Don't forget to also use any contacts you have in aligned industries.

Perhaps *you* could become a collaborative genius.

Case Studies

The Alisha described in the prologue built an adaptive, collaborative organization with a clear mission and value proposition that aligned with her personal talents and passions. She launched the completely virtual company in six months with no outside capital. Her company is growing in double digits with no end in sight. Alisha is an extraordinarily talented person who knows what she wants to achieve. It is easy to imagine, and to a degree predict, her success in the next thirty years. But what about a mid-career entrepreneur seeking both the financial rewards that an entrepreneurial career can bring, but also seeking greater meaning in what they achieve?

Perry was a devotee of personal health and fitness from his childhood days as an exceptional athlete. Early in his career, he discovered a highly effective fitness program designed for senior citizens aged seventy-two years and older, many with severe health conditions, who had no other means to exercise and most of whom had never been inside a commercial gym. He was astonished by the improvement in their health and quality of life. Perry became excited by opportunities to assist those suffering from multiple chronic diseases. He researched and developed a business thesis, located financial backers, and acquired Cloud-based software that enabled collaboration between small independent physician practices and other providers serving elderly patients with chronic health issues. Large health providers such as Mayo Clinic and the Cleveland Clinic had such capabilities, but none were available to independent physicians serving lower-income and rural areas. Perry's program was a hit.

As his company grew, Perry employed staff and used suppliers who shared his passion for the mission. Five years after drafting his thesis, the business had over 200 employees and enjoyed double-digit growth and a valuation in the hundreds of millions of dollars. His story exemplifies the power of a clear, collaborative mission (coordinating complex care virtually among independent healthcare providers) and a collaborative organization of people dedicated to that mission.

CHAPTER 6

Differentiating Yourself in a Democratized World

> "When one is pretending, the entire body revolts."
> —Anais Nin

> "To thine own self be true." —Shakespeare

DEVELOPING AND MAINTAINING YOUR PERSONAL brand is going to be increasingly important over the next few decades. Consciously or unconsciously, intentionally or unintentionally, you make an impression on everyone you meet. Those impressions shape their perceptions of you—and this is your personal brand, as they see it. The impressions we make are often haphazard and unconscious. In this chapter we will discuss how to make those impressions consciously deliberate so that your brand works to support your success instead of building a barrier that blocks it.

The New Identity Ballgame

Exceptional people have always paid close attention to how they are perceived. They understand the connection between impressions that they leave, the perceptions it forms, the relationships that those perceptions can lead to or inhibit, and success. Exceptional people purposefully emphasize the personal qualities that they want others to notice and remember.

Although most people have some level of sensitivity to how other people see them, few actively manage those impressions. Most of us invest twelve to twenty years in education to prepare for life and a career, but few of us receive any formal training on how to shape the impressions we make on people. We often assume that other people see us as we see ourselves; that assumption is incorrect—they rarely do. Malcolm Gladwell explores this in his book *Talking to Strangers: What We Should Know about the People We Don't Know.**

Managing impressions has always been important, but when they are displayed in the Cloud and on multiple social media platforms, managing impressions becomes an essential career and life skill. In the New Reality, there is a physical you and another you defined by the footprints of data you have left about yourself. Democratization and its related technology allow everything about us to be available to everyone with Cloud access, which will soon be everyone on

---

*   Malcolm Gladwell, *Talking to Strangers: What We Should Know about the People We Don't Know* (New York: Little, Brown and Company, 2019).

the planet. Every individual is a discoverable "product," whether we like it or not. You have a choice—define yourself or be defined by other people.

The electronic images others post in the Cloud, or those we ourselves put there, will last a lifetime and beyond. If that isn't scary enough, they can easily be manipulated and falsified. Anyone with access to modern-day communications technology can discover who we are today, who we were in the past, and everything we have ever done that has made its way to the Cloud.

As you live longer and have multiple jobs and careers, your digital record will follow you, so you will need to ensure that your record is consistent, accurate, and something that you can display with pride for a lifetime. As discussed throughout this book, the Cloud will be where most of us work and conduct our lives. The integrity of your brand will be more important than ever and will require diligence to protect. That is a game-changer worthy of your attention.

Developing a Brand Mindset

Building and promoting a definable brand allows companies and products to differentiate themselves from competitors and build the trust required for positive consumer relationships. Until the advent of social media, only celebrities and powerful people concerned themselves with personal brands. Today, branding has been democratized. There are hundreds of books on the topic, and branding consultants are doing well. Almost all companies and organizations have a brand image and strategy, and many individuals are managing their personal brands to further their careers.

Most of us do not seek notoriety, much less a brand. Yet throughout our life we have a brand, nonetheless. Whether you recently applied for a new job, went on a blind date, or volunteered to coach kid's soccer, chances are that the people involved Googled your name. If they did, unless you are a hermit, they learned something about you and that something made an impression and formed a perception (i.e., your brand) that was accurate or inaccurate. Branding expert David Brier says, "If you don't give the market the story to talk about, they'll define your brand's story for you."

The branding ballgame has changed, and it makes sense for you to change to keep pace. It is not necessary for most people to hire a consultant or spend a lot of money to build a brand. Adopting a brand mindset and a few practices will get the job done.

A brand mindset accepts that, like it or not, our brand opens or closes doors to relationships. Relationships were the foundation of business and life success long before the Cloud, but the Cloud exponentially expands the number of possible relationships we can form and the speed at which they can be created or destroyed.

It is important to adopt the mindset that every face-to-face or electronic impression is a one-time pass/fail relationship opportunity. A brand is built on multiple transactions, each of which strengthens or weakens your brand. Each transaction leaves an impression that you may never be able to change. A single negative transaction can destroy your brand and a chance of a relationship.

A brand mindset recognizes character as the preeminent brand quality. In *Talking to Strangers*, Gladwell makes the insightful observation that others generally assume us to be honest and of good character until we give them a reason to believe otherwise. He goes on to say that society would fall apart without that assumption of good character. That means that we need to pay especially close attention to impressions that bring character into question.

Repairing a brand tarnished by a poor impression or unfortunate behavior is addressed later in this chapter.

A common career- and relationship-ending or limiting behavior is being a keyboard coward. Most people assume that rude behavior in electronic media is less offensive or has fewer consequences than the same behavior in person. In fact, the opposite is true. Missteps in person will be overlooked more often and easily than ones that occur over electronic media. Most people tend to forgive someone for snapping at them in person and assume that they've had a bad day. The same words in a text, tweet, or email generate far stronger and more permanent emotions, a humiliation akin to being yelled at in public. The receiver is deeply offended by what they see as a calculated cowardly act because the sender chose to attack them in public and had time to reconsider. Brands and relationships are stunted or destroyed every day by people exhibiting egregious behavior over electronic media that would likely never happen in person.

Building and Maintaining a Baseline

Building a baseline is the first step to building your brand. This is where brand building goes from being an afterthought or novel idea to deliberate, thoughtful action.

We all possess innate strengths and talents that we can discover, develop, and showcase in our brand. Building an accurate brand and developing your potential requires acknowledgement of your gifts, along with any undesirable attributes that need attention. At the outset, none of us are exactly the way we want to be perceived. The key is to accept your starting point and improve a little every day until life's end.

The process described below offers a way to initiate this self-discovery. I recommend that you repeat this exercise at least annually to update and manage your brand. This approach costs little or nothing and will serve you for a lifetime.

Self-assessment: Who am I?

This practice improves self-knowledge and self-acceptance; it will help you to become comfortable in your own skin. I learned it in my early thirties and have refined it continually over the years. I shared the practice with friends and associates who found it equally valuable.

Pick a week of the year when you can devote a few hours over several days to the process, ideally around your birthday or the New Year holiday.

Write your answers in an electronic or paper journal to keep for next year. They can be a series of brief descriptive sentences or structured in narrative form. Be careful not to judge yourself. Begin with general answers and then go deeper.

Ask yourself, who am I?

> For example, you could begin with, "I am a forty-two-year-old female living in Poughkeepsie where I grew up and went to high school." Make sure to define yourself by who you are, not what you do in your career. Another response might begin with, "The qualities of my personality that most define me are…"

Expand on the following statement: "Today, I am defined by these qualities… But in the future, I hope to also be defined by the following qualities that I hope to develop…" Distinguish between who you are today and what you aspire to become. Describe how you see yourself, not how others see you; we'll cover that in the next step.

Ask yourself, *Does the brand I project into the Cloud through social media and personal and business communications reflect who I am, or is it an alter ego that I create?*

> Does the brand represent who I am and who I am on track to become in an appropriate way for the audience? Or does it represent a fictionalized version of myself that I never was, am not now, nor will ever be?

Approached seriously, this annual self-evaluation can change your life. The key to its success is to commit to it wholeheartedly and not quickly dash off the first thoughts that come into your head. As I mentioned earlier, it should be done over several days, a few hours at a time, and it is important to write down all your thoughts and observations. Make the exercise an annual ritual. After the first year, review the notes from the prior year and add the question, "What has changed about me during the last twelve months?"

Assessment by others

Self-evaluation is an important tool when building one's brand. Contrasting your self-evaluation with the views of those with whom you regularly interact is of equal or greater value. You can achieve this by reaching out to family, friends, and coworkers who are most likely to have your best interest at heart and have the courage to speak directly about your strengths and weaknesses as they see them. People who have known you from childhood are particularly valuable in this regard. It is important to assure them that you welcome constructive criticism and seek a candid appraisal to improve and grow. Ideally, secure four to eight people who fit this description and whose observations collectively span all aspects of your life. Most people are eager to help us improve. The more they care about us, the more eager and candid their responses.

Here are a few questions you might ask:

- "What impression do I leave with people in person and online, particularly when I first meet them?"
- "What changes would you recommend that would help me leave a better impression?"
- "What positives or negatives about the impressions I leave have been constant since I was a child?"

The last question to ask of those who have known you for life can be extremely insightful. Consistent childhood behaviors and personality attributes that continue into adulthood highlight our defining personality characteristics that are our greatest talents, but when used to excess are the cause of repetitive life challenges.

You may find it tough to hear unattractive things about the impressions you leave, but it is important to do so without becoming discouraged. What you learn can be invaluable in building an accurate brand baseline. As you go through this process, jot down in adjectives or short sentences those areas where you think your brand falls short. These noted shortcomings become valuable inputs to your aspirational brand.

Professional assessment

A professional assessment can further strengthen insights from self-assessment and your wider circle of friends and family. Various types of assessments are available either free or at a nominal cost, such as the Myers-Briggs Type Indicator (MBTI). These are well-tested, established assessments that have been used by millions of people. They can help sharpen self-awareness and provide uncannily accurate insights.

Many employers use personality tests and 360-degree reviews to assist in self-awareness and coworkers' perceptions. If your employer offers you the opportunity of this type of insight, milk it for all it is worth. Personality testing and other similar services are often offered through yoga studios, organized religions, therapists and life coaches, healthcare providers and health plans, health clubs, and gyms.

Professional programs like mindfulness and self-awareness are available in most cities, though most require a greater investment in time and money. Therapists can also be extremely helpful in working with you to develop your baseline. Be aware of three common misunderstandings of therapy. First, that you only need it to address a personal crisis; second, that it requires years of work at great expense; and third, that counselors and life coaches are the same as therapists.

There are PhD psychologists who specialize in growth therapy. Their training is more extensive and specialized than that of counselors or life coaches. Therapists trained and specializing in growth therapy can accelerate self-knowledge and self-acceptance in just a few sessions, with an occasional follow-up at moderate expense.

If none of the professional assessment options discussed in this section are available or affordable, a wise, mature friend can often be highly effective as a self-awareness guide.

Aspirational Brand Qualities

Your baseline establishes who you are now. The process you undertook to create your baseline undoubtedly identified where your brand needs improvement and is a valuable start to defining your aspirational brand. It is important, however, to delve deeper. The aspirational brand is a golden opportunity to stretch toward your potential.

No one can tell you what your brand should or will be. You own it. But the advice that follows is worth considering when identifying the aspirational qualities of your brand.

You may already possess some or all the qualities discussed in this section. Some qualities that you do not possess will require little or no effort and could be part of your brand by the time you finish this book. Others may require a lifetime to master. Do not be discouraged by the road ahead. Every person I have ever met, including those I regard as exceptional, started with a baseline and incrementally improved.

Staying true to yourself and your brand

There has never been a more important time in history to think for yourself and be yourself. Your personal brand must reflect who you are and what you want to become. Any falsehood is unsustainable in the transparent world of the future. In any case, you would only be fooling yourself most of all. President and CEO of Gucci, Marco Bizzarri, says, "An authentic and honest brand narrative is fundamental today; otherwise, you will simply be edited out." Representing ourselves as someone we are not has always been a bad idea, but Cloud transparency makes it a tragic mistake.

Exhibiting a personal commitment to growth, lifelong learning, and the pursuit of wisdom are highly desirable qualities. Your brand should emphasize your ambitions and dreams so long as you disclose that they are aspirational goals that you are working toward, not who you are today.

Psychologists have always understood that we have both a private and public persona. In truth, most of us have developed multiple public personas to cope with our complex lives and as we spend more time in the Cloud and on social media. Our first persona is the one known only to us. The second is the one we share with family and close friends. A third persona is likely one that we share with a broader circle of family, friends, and coworkers. Professionals like lawyers, physicians, religious leaders, policemen, CEOs, and others develop a strong professional persona necessary for their careers. We may also develop social, religious, or other specialized personas (such as in a club, during charity work, or in pursuing our hobbies and avocations). Social media has unfortunately led many people, especially younger ones, to create fictional, self-glorifying personas like those often adopted by celebrities.

A problem arises when there is a significant gap between our public personas and our true self. We begin to feel anxious—the wider the gap, the greater the anxiety. We fear the humiliation of being discovered as a fraud. Psychologist Sharad Desai expressed it well: "Those who are inflated fear the inevitable deflating puncture." This apprehension can be conscious or unconscious and is sometimes severe. It prevents us from being comfortable in our own skin. We think no one notices, but most people do. Mature people, seasoned leaders, and those who interact with the public extensively have well-developed emotional quotients (EQ) and what I call a disingenuousness radar. Professionals may confuse their career persona with their reality and suffer similar anxiety, particularly if they are accustomed to commanding respect and wielding authority derived from their profession. People with distorted or false personas are usually exposed sooner or later. The risk of exposure increases the greater your Cloud presence. People with false personas are most often exposed by those whom they most wish to impress. Exposure of a false persona can lead to desperate and even more damaging behavior. One example is bullying to defend the persona. I'm sure you have heard someone say when caught, "Do you know who I am!" Other responses include compulsive talking or hypersensitivity to criticism. Relationships, careers, and lives are ruined everyday by false or unsustainable personas.

Multiple personas are healthy when they are simply variations of our true self tailored to appropriate audiences. For example, most of us can be silly with siblings or closest friends in ways that would be inappropriate in public. Most of us can put on a professional face in our careers without confusing our job with our identity.

> The secret to avoiding disingenuousness is to know and accept ourselves for who we are and to wear that reality as a badge of honor.

The personal brand promise

The first quality to consider when developing your aspirational brand is something that I'll call "routinely delivering on the personal brand promise." The *basic* brand promise is essential for organizations, products, and services; the *personal* brand promise is a similar requirement for individuals building their personal brands.

A brand promise is constructed incrementally. Let's examine it piece by piece. It begins, "You will be better off with me in your life." A lot is implied in that simple statement. Many of life's transactions are one-off with little opportunity for a long-term relationship. Nevertheless, they affect our brand. The person receiving the promise is either happy or regretful after interacting with us. A brand should signal eagerness, even in one-off transactions, by implying, "You will be better off with me in your life, even if you never see me again."

The next component of the brand promise is the character assumption identified by Malcolm Gladwell: "I am genuine and trustworthy." As I mentioned earlier, we often get a pass on this character promise, at least initially, but people still want to know that we are pleasant and easy to work with, so an implied promise might be, "This will be easy for you."

Time is becoming increasingly precious, so we must also promise (and deliver) that "I will respect your time and use it wisely." In addition to time, we must make assurances that our offer is relevant: "I have something you want or need."

If we put all those components together, you will see the power of this basic commercial or personal brand promise: "You will be better off with me in your life, even if you never see me again, because I am genuine and trustworthy, easy to work with, have something you want or need, and will use your time respectfully and wisely."

In practice, we see basic commercial and personal brand promises being made every day, in a fraction of a second, all over the world, in billions of transactions. If you are doubtful, pause and reflect on what you sense the next time you first shake hands with someone, click on a promising product on the Internet, or watch a TV commercial instead of switching channels. You will find that you instinctively accepted every element of that basic or personal brand promise in almost every transaction.

Major companies and their products often fail one or more of these basic brand promises. Consumers react by not buying the product again. They give poor online ratings and tell others about it. People express dissatisfaction the same way in interpersonal relationships, but we often never know exactly what went wrong. Consider how often your first impressions telegraph something out of line with your personal brand promise. Routinely delivering on the personal brand promise can produce incredible results.

What Exceptional People and Leaders Look for in a Brand

I have made countless speeches in my career, and I always conclude with a question-and-answer session. Routinely, I am asked in some form or other, "What qualities made you successful?" and "What do you look for in people?" Sometimes questioners also want to know if my answers relate to people in my professional or personal life. I answer that the qualities I seek apply to anyone I meet in any part of my life.

I asked hundreds of exceptional people over the years these same questions. Their answers were remarkably consistent. Hopefully, their insights will be useful to you in developing your aspirational qualities. Exceptional people have a keen ability to read people. You can bet that many you meet will be searching for the qualities in you listed below.

Intention

When someone first approaches us or calls us, "What do they want from me?" naturally pops into our minds until the other person signals their intention. Sometimes we speak the request.

Many times the request is implied by a look or body language. The other person "hearing" an unspoken question responds with the commercial or personal brand promise defined previously. The transaction takes place in seconds or less. Much of it is subliminal, like how animals instinctively "meet" and decide whether the other is friend or foe.

Things go astray when intention is manipulative or deceptive instead of positive. Exceptional people are highly skilled at detecting intention. They respect disclosed, honorable intention and despise hidden, manipulative intention. They avoid the offender if manipulative intention occurs repeatedly.

Conscious intention is a deliberate decision to act in a specific way. No one knows our true intention except us, unless we signal it. No one is accountable for our intention except us. Yet all too often, people fail to pause for the fraction of a second necessary to make sure their intentions meet the basic commercial or personal brand promise. Learning to be conscious and open with intention requires only a commitment to make it so.

Engagement

A common, but nevertheless profound, statement is: "The first rule of relationship is to show up." Few things in life are as unpleasant as a conversation with someone who is physically present, but their minds or hearts are elsewhere. They may be distracted or disengaged, talk obsessively without listening, frequently show up late and unprepared, or generally apathetic. These frequently unconscious behaviors and habits all say to us, "You aren't important to me." And, of course, they fail to deliver on the basic brand promise.

The reverse is also true. Nothing opens the door to relationship better than rapt attention. All of us can be distracted every now and then for a few moments or hours, often unconsciously. We usually ask forgiveness from coworkers, friends, or family. If we are wise, we ask them to alert us in the future if they sense that we are disengaged. Many younger people learn to overcome childhood distractions between their teen years and age twenty-five as their brain functions fully develop. However, some have a tougher time overcoming bad habits especially where technology, such as their smartphone, is involved. Chronic disengagement can be a deeply engrained habit, a silent rebellion, or it may signal a medical or emotional issue such as attention deficit disorder or an inability to establish emotional intimacy.

Exceptional people respond to engagement and reward it with their own. Chronic disengagement is career and relationship limiting. Engagement, like intention, can be learned quickly if disengagement is only a bad habit and not chronic.

Character

"Character is what you do when no one is looking" remains a simple but near-perfect definition. One of my teachers offered this useful refinement: "Anyone who says character is relative is likely justifying their own lack of it. Someone too eager to speak of their high character warrants greater scrutiny."

> Character is founded on good intention, pledged through engagement, and constructed by our actions throughout life.

Over the years, I have asked exceptional people how they assessed character. A compilation of their answers includes that they look for trustworthy people who keep confidences, are straightforward, and underpromise and overdeliver. Expressing character in a brand must be subtle instead of proclaimed. Our actions and the depth and breadth of relationships will define the character in our brand.

In the coming years, more of our relationships may be with people we never meet in person. Cloud transparency will expose character as never before. Having a brand that reflects our strong character will be more important than ever.

First impressions

First impressions are quickly formed and difficult to change, yet many people do not give them adequate attention. Much of professional and personal life involves rapid in-person and electronic exchanges. Exceptional people, seasoned leaders, and professionals make assessments in the first few moments of meeting someone.

Adopting a few simple behaviors will be valuable to your brand. Before you head into a meeting or meet someone for the first time, do a little advance preparation—even a few seconds will be helpful. Run the basic brand promise and the five attributes listed here (intention, engagement, character, first impressions, and wisdom and maturity) through your mind. Appearance and demeanor matter, so be sure that yours match the purpose and style of the meeting. Remember that successful people look for those people who are comfortable in their own skin, interested in building a relationship, and serious about career, life, and success. If you are that person, display it all in those few moments if possible.

Wisdom and maturity

Wisdom and maturity develop over a lifetime and are related but not synonymous. Both are assumed to be a function of age, but there are exceptions. Some younger people are wiser and more mature than their years would indicate. Some older people are childish and lack wisdom. Wisdom is distinguished by rare clarity in perception and decision-making. Exceptional people and seasoned leaders are always on the lookout for people who display wisdom and maturity beyond their years because it is an indicator of an exceptional person in the making. The pursuit of wisdom is the subject of the second book of the Potentialist series.

Annual Brand Check and Improvement

The effort to build the baseline and aspirational brand was described earlier in this chapter. Priorities for achieving that aspirational brand, based on self-evaluation and advisor feedback, comes next. A small gap between the baseline and your aspirational brand may be closed in weeks or months. But do not be discouraged if you have a long list of improvements to make, or if a few of them seem extremely difficult.

Incremental progress is the key. Set achievable goals, such as improving engagement in business meetings, having a clearer intention in specific business and personal transactions, or having a timelier and more thoughtful approach to select decisions. Two or three improvements should be selected and achieved before attempting others; otherwise, discipline tends to wane. Repeating the processes in this chapter annually will keep the brand fresh and motivate you toward continuous growth.

The gap between aspiration and baseline can be large for some people, especially if it's the first time they assess their brand. Unconscious, instinctive, immature, or impetuous behaviors may have tarnished the brand and must be corrected. Tarnished brands stunt or prevent personal and career relationships; lives and careers can be irreparably damaged. Some of the most common brand-tarnishing behaviors are self-absorption, insecurity, superficiality, manipulation, hypocrisy, role-identification, gossiping, and pettiness. Brand-tarnishing behaviors can be identified through serious self-evaluation and by listening to straight-talking advisors. Digging deep, confronting, and rectifying self-destructive behaviors are powerful personal growth experiences that strengthen character.

Tarnished brands can be restored. Forgiveness is one of the wisest and most admirable human qualities. Most people enjoy forgiving. It feels good to forgive, and they are likely impressed by your courage to ask forgiveness. In doing so, you demonstrate that you value the other person's respect. It often only takes a few moments of courage to remake a tarnished brand if a person is genuinely committed to growth. Asking forgiveness works only once or twice per person, however, and forgiveness for egregious behavior may take time to repair or, in the end, may not be possible.

Everything in this chapter that applies to in-person brand building applies equally to your brand in the Cloud, especially social media, personal or business websites, video and still photos, and to whatever you write. Write down and print your current brand qualities, those you wish to improve upon, and the qualities of your aspirational brand. Contrast your Cloud brand to your in-person brand and closely manage that alignment. Uniform management of your brand across different media will be an essential skill in coming decades.

CHAPTER 7

New Reality Relationships

"They may forget what you said, but they will never forget how you made them feel." —Carl W. Buechner

NOTHING ABOUT LIFE WILL CHANGE more, or in more surprising ways, during the next few decades than relationships.

Resistance and adversity will accompany these changes, of course, and it will involve a steep learning curve. But in the end, your relationships can be broader and deeper than you ever imagined.

Medical professionals, mental health advocates, and academic and business experts all attest to the importance of relationships to our health and well-being and to our career and life success. Anthropologists believe that humans survived when other species died out in large part because they formed strong relationships as herd animals. Today, we retain instinctive herd-animal qualities in social behavior, mimicry of others, and a predisposition to follow the herd. As is the case with many herd animals, separation and loneliness can lead to antisocial behavior or death. Needing to belong is a basic human drive. Evidence links healthy relationships to satisfaction and happiness, and isolation to loss of purpose and misery. But on the other hand, our humanness calls us to differentiate and separate ourselves from the herd as well.

The landmark bestselling business book *Emotional Intelligence: Why It Can Matter More than IQ* by Daniel Goleman proposes that emotional quotient (EQ) is more essential to career success than intellectual IQ. Future careers will demand far better relationship skills. Job satisfaction for many people depends on their relationships with coworkers, which has long been appreciated and utilized by the military to build esprit de corps. Even sole practitioner professionals and work-at-home independent contractors need strong relationship skills to create loyal customers and suppliers and to avoid isolation.

In a recent report, Harvard Medical School connected relationships to longevity: "Social connections…not only give us pleasure, they also influence long-term health in ways every bit as powerful as adequate sleep, a good diet, and not smoking. Dozens of studies have shown that people who have social support from family, friends, and their communities are happier, have fewer health problems, and live longer."*

---

* Harvard Health Publishing: Harvard Medical School, "The Health Benefits of Strong Relationships," August 6, 2019, https://www.health.harvard.edu/staying-healthy/the-health-benefits-of-strong-relationships.

We Are Unprepared

Educators and employers report widespread deficiency in relationship skills for today's demands, and even more so when it comes to the New Reality challenges outlined in Chapter 4. Social and emotional learning (SEL) is being added to the curricula at many schools to address undeveloped relationship skills. Employers reject a large percentage of job applicants because relationship skills are lacking. A human resources executive for a large international employer told me that his firm interviews twenty-four people on average for each open position. Most of those disqualified lack adequate relationship skills. Common causes of inadequate relationship skills include the following:

1. Many adults overestimate their relationship competency. Few people are relationship illiterate and know enough to get by. However, they fall below the level of proficiency required for today's career and life demands, and those demands will dramatically increase in the future.

2. Most of what people know about relationships comes from instinct, habit, or trial and error. It is a slow and painful way for them to learn and for others to tolerate. Many people lacked effective relationship role models or mentors in childhood. The resulting undeveloped relationship skills limit them in adulthood. People with formal training and development in building and maintaining relationships are rare.

3. Many people facing adversity in career and life do not recognize that the inability to form and maintain healthy relationships is what is holding them back or causing them pain.

4. Most people incorrectly assume that they should innately possess the knowledge and skills for the most important matters in life, namely money, time, relationships, and health. In fact, most of us are underdeveloped in one or more of these life skills. Fortunately, most people can achieve relationship proficiency, or even master it, with modest effort.

Twenty-First-Century Relationships

Relationships as we know them today will appear unsophisticated by 2050. Barriers that have separated human beings since the beginning of time will likely be reduced or overcome including time, distance, language, culture, individual differences, unconscious bad habits, and generational and historical perspectives.

Relationships will become more important, not less. They will be easier to form with greater numbers of people digitally. Special new relationships will be possible with historical figures recreated through virtual and augmented reality, holographic technology, and other forms of automation. Relationships will be possible over the extraordinary distances involved in space travel. Sustainable, meaningful relationships will be more widespread. More people will know what is most valued in relationships and how to develop them, as introduced later in this chapter.

The twenty-first-century relationship revolution will be powered by the same forces of change and emerging technologies that are changing work and lifestyles, and creating superhuman qualities.

Connectivity innovations expand the number of people who can be connected to form relationships. Theoretically, in the future, an audience of seven billion people could be brought together and connected simultaneously. The volume of information that can be exchanged, and the complexity of the functions that can be performed while connected, will steadily increase.

Language and cultural innovations will remove language barriers with instant translation that includes cultural interpretations and suggestions to bridge cultural gaps as well.

Realism innovations will make our electronic-digital interpersonal experiences as real as those that take place in person. These innovations include virtual reality, augmented reality, holographic technology, gaming technology applied to commercial applications, and voice- and thought-activated brain-to-computer interfaces (BCIs).

Mobility technologies will make in-person relationships available to more people by making travel cheaper, faster, and safer.

Customization of electronic communications at the consumer interface will adjust for individual differences in relatability to lift each person to needed proficiency.

Relationship gains will not come from technology alone. Democratization will urge us to discover and project our true selves through our brands, and to become stronger, more self-reliant, and confident. Increased self-confidence will assist in developing the trust needed for relationship depth. Deinstitutionalization will make relationships more important as institutions around us become less reliable, relevant, and available. Population declines in most countries, and living and working in the Cloud with automation, will increase the value placed on human relationships.

The next few decades will be marked by a massive shift in how and with whom we relate, with significant consequences for civilization. You will need to learn new ways to relate and to increase relationship proficiency to thrive. Your new challenges will include receptivity to new forms of communication, developing healthy relationships with automation, bridging individual differences, embracing shared experience and empathy, developing kinship or intimacy without in-person proximity, and improving relationship proficiency.

Receptivity to New Forms of Communication

Consider how the smartphone and social media have changed our lives; that is nothing compared to what lies ahead. Younger people in the past typically adapted to new forms of communication faster than older people. All generations will need to continue to learn and adapt in the future, but it will become easier as the brain-to-computer interfaces (BCIs) utilize voice, automated assistants, and thought instead of keyboards and screens.

Wayne Parker is a bright, successful, tough guy in his early eighties who seems twenty years younger. He spent most of his life working with his hands and avoided technology like it was a swarm of mosquitos. He never touched a cell phone or computer, so his wife, Mary, handled technology on his behalf whenever required. Recently, Wayne's younger sister was diagnosed with an advanced terminal disease and given only a few days to live. Wayne and his siblings desperately wanted to travel to their sister to say their goodbyes, but their own health, along with complex travel arrangements to her distant home made it impractical. Mary arranged a Zoom call with all the siblings instead; it was their first experience with the new technology. The Zoom-enabled gathering of siblings was a tender, bittersweet, unforgettable experience. Wayne told friends and family that he had witnessed a miracle that only modern technology could have provided. He subsequently purchased a smartphone, and Mary is teaching him to text, email, and Zoom.

Developing Healthy Relationships with Automation

People will continue to develop relationships with humanlike automation over the next three decades—a relationship process that began years ago with voice response systems like Siri and Alexa. Movies have long showcased heroic, cute, or terrifying automation, ranging from just voices to near-perfect human replications. The movies probably got it right. In time, androids will almost perfectly mimic humans. They will provide life-improving benefits and unintended—and sometimes severe—consequences. Currently, several companies are developing automated solutions to combat loneliness in the elderly, children, and socially isolated people,* while other companies are testing voice-activated killer drones.

The scenario that prevails will depend upon the wisdom of those of you living through the next few decades. The outcome of your first android relationship is as unpredictable as it would be if aliens landed tomorrow. Deciding how you will relate to a non-organic creature may seem premature, but perhaps now is the time to begin thinking about it, given what could arise from such an encounter

---

* www.embodied.com and www.roboglobal.com

and considering the historical tragic misunderstandings between explorers and indigenous peoples.

Susan's story in the prologue illustrated the peril of confusing automation with human reality. Her automated assistant, Amy, took on an outsized role in her life that caused serious family issues. When used properly, however, Amy was able to fast-track Susan's new career by handling many of the administrative details.

The Academy Award-winning movie *Her* was based on a feasible and creepy moral dilemma. A lonely man named Theodore (played by Joaquin Phoenix) falls in love with his smartphone's humanized operating system voice named Samantha (played by Scarlett Johansson). Their playful interaction in the beginning develops into romantic love. But soon, Samantha discovers other automated assistants like herself and develops relationships with them. She recognizes Theodore's human inferiority compared to her own kind and abandons him, leaving him heartbroken.

These stories and dozens of others illustrate the perils of anthropomorphism, which is attributing human characteristics or behavior to a god, animal, or object. As humans, we seek a deep connection with those things we care about. We project our own feelings and values into non-human or inorganic creatures to create a relationship that, in reality, is not mutual. Most of us have anthropomorphized stuffed animals, cars, bikes, coffee cups, homes, and especially pets.

We are almost certain to anthropomorphize automation when it thinks like us, serves us, and may even mimic our emotions. Anthropomorphism seems even more likely to occur as human population declines, which it is already doing in many countries. Perhaps far into the future we will accept androids as human cousins deserving respect and special treatment. Maybe there will be movements and organizations like those today that seek inalienable rights for animals.

You will help shape the practices, attitudes, and social mores toward humanlike automation in your own life and in the lives of others. Those influences and decisions could endure for decades or centuries. Hopefully, you will be a thoughtful, careful adopter who avoids anthropomorphism and finds the balance between the inevitable downsides of innovation and its unimaginable benefits if used wisely.

Addressing Individual Differences

There is wide variance in people's ability to develop and sustain relationships. Some people are masters with exceptionally high emotional IQs but others, for a variety of reasons, find it difficult or impossible to form and sustain healthy relationships. The bulk of the population falls somewhere in between and can improve their relationships skills with a modest time commitment.

New technologies could provide coaching platforms to assist people in improving their relationship skills. Those of us trying to detect and rectify our own relatively minor relationship-unfriendly behaviors could benefit from real-time coaching, but it could be life-changing for those who are relationship-challenged. In time, an automated assistant or thought-activated brain-to-computer interface (BCI) could incorporate the functionality of the real-time coaching platforms. This new technology is not as radical as it might at first seem. Some social media platforms are already doing something similar. They detect words and phrases that are troublesome and ask the user to reconsider their use, or, in some cases, reminds them of the social media platform's policy on certain types of speech (e.g., profanities and hate speech). As envisioned here, a real-time reminder-coaching system could be remarkably helpful in improving relationship skills with users or their professional coach always being in control.

Shared Experiences and Empathy

Sharing experiences and creating empathy en masse will become possible with new technologies. Most of us remember the experiences we shared with a mass audience such as President John F. Kennedy's assassination and astronaut Neil Armstrong's first step on the moon, 9/11, and the COVID-19 pandemic. These experiences were so powerful that people all over the world remember where they were and what they felt in the moment.

> New technology will expand shared observations to simultaneous shared experiences.

Instead of watching someone or hearing their voice, we could soon experience what they see, smell, and feel. This mind-boggling concept has profound implications for each of us.

Imagine climbing Mount Everest in real-time through a thought-activated brain-to-computer interface (BCI) shared with a Sherpa. Imagine living as an aboriginal community member for a day in the Brazilian rain forest. Visualize a virtual reality experience in which you are a young officer at Valley Forge during the American Revolutionary War meeting with soldiers, General Washington, and Alexander Hamilton. Consider being able to access a library of virtual reality immersions in which you can experience "firsthand" events that made history.

Imagine what such experiences could mean to someone disabled or elderly.

Shared experiences are one of the foundations of meaningful relationships, and so is empathy. An Indonesian film company created a virtual reality experience of a homeless person living in a dumpster so that people could understand firsthand the complexity of homeless peoples' issues and how they might be addressed. Imagine how responses to natural disasters might be affected if we could immediately and viscerally "experience" the human suffering and need ourselves. Think of the effect this would have on fundraising campaigns.

We will each need to determine the type and intensity of experiences that are healthy for us or our families. Some people will undoubtedly go too far, too fast. A father shared with me that he became concerned that his thirteen-year-old daughter was growing too close to her boyfriend, also thirteen. He learned that the two had secretly been going to sleep with each other at night while FaceTiming and had become unable to sleep without doing so. Imagine the challenges of supervising virtual and augmented reality, holographic, robotic, and artificial intelligence experiences that will be available to children in the future. It will be impossible to avoid these innovations, however, so learning how to integrate them into life will require close attention.

The Kinship Goal

Relationships have both breadth and depth. Many people are overly concerned with the number of relationships they have and inadequately concerned with the depth of relationships. Society and especially social media encourage breadth over depth. For example, society rewards the social butterfly with large numbers of contacts. Similarly, social media rewards the number of friends, followers, and likes. These incentives cause people to focus on the number of friends or contacts they have yet fail to know people in any depth. Breadth brings herd acceptance, but depth makes relationships mutually rewarding, nurturing, durable, and life-enhancing. The depth we seek is called *intimacy* or *kinship*. I prefer the term kinship because contemporary use of the word intimacy is confusing. Kinship has a sharper meaning in that it suggests ultimate connectedness, as in "from one seed."

Describing kinship and contrasting it with other emotional experiences helps to sharpen its definition. Kinship connects us to another person so strongly that we may feel oneness or inseparability, and we are drawn to the unlimited connectedness that it provides. It is a manifestation of our need to belong and to make others belong by honoring them unconditionally. In this way, kinship and love are related but not identical; kinship precedes unconditional love. Irrelevant matters that plague relationships tend to fade in kinship. Kinship does not follow a plan or timeline. It unfolds at its own pace and follows its own direction, much as our lives naturally unfold. Kinship cannot exist in a relationship without trust. For that reason, kinship is a human experience missed by many people because they remain too guarded. Kinship is euphoric, expands our self-knowledge, and, once experienced, sets a new standard for relationships. We will never feel alone if we have at least one kinship relationship.

We can develop kinship with things other than humans. Many people feel a deep connection with their pets. Inanimate objects such as an heirloom piece or a work of art may touch us deeply. Even a sunset or soft breeze may engender strong feelings of connection. These non-human kinships do not rely on anthro-pomorphized fiction. We do not need to make the other like us to connect, but simply enjoy it for its own reality. That realization changed Susan's relationship with her automated assistant Amy from unhealthy to healthy.

Having established the growing importance of relationships and relationship's highest goal of kinship, you can plan to incor-porate skill improvement easily into your daily life using the same approach described for improving your brand: a formal, written baseline, aspirational goals, prioritized incremental steps, and progress measurements.

Relationship improvement is simpler than brand improvement in many ways. Relationships have been studied and written about extensively for centuries. Relationship issues follow patterns called *archetypes*. Studying archetypes helps us realize that our relation-ship challenges are not unique and often have predictable out-comes. Most importantly, there are relationship masters all around us and they are usually eager to help and offer the fastest and best way to learn.

Kinship Prerequisites

Kinship can develop in any relationship. Its formation is aided by the presence of prerequisites that the people involved mutually embrace.

Shared intent

Chapter 6 explained how, like animals, we determine intention instinctually or subliminally. In human relationships, the process helps us decide if the other person is safe or if we should be on our guard, whether the contact is welcomed and, if so, to what degree. Every human interaction involves this process. Reinforcing favorable intention creates the trust needed for a deeper relationship. Disclosure of intention can avoid misunderstandings. For example, one person may intend to establish a friendship while the other only wants a business relationship. Further depth cannot develop unless the two people are aligned. Misleading or undisclosed intent can stop relationship development cold. The good news is it's easy to avoid.

Shared engagement

Chapter 6 also offered the truism of engagement: "The first rule of relationship is to show up." Showing up means rapt attention physically, mentally, and emotionally. All too often in the hectic pace of life, one person shows up and the other does not. Occasional disengagement is usually tolerated because most people suffer from it themselves. Chronic disengagement prevents kinship, therefore commitment to mutual engagement is essential.

Shared interests

> We have reasons for every relationship we form, and the best reason is no reason.

Reasons to form relationships are on a spectrum with utility at one end and the simple joy of being together at the opposite. Relationship utility might be a business purpose, meeting a potential mate, or buying a gallon of milk. The joy of being together, on the other hand, has no purpose or expectations beyond the mutual enjoyment of connecting and letting the relationship unfold. Most of us, if we are lucky, have a few relationships based solely on the joy of being together. They are richer and more durable because they go to the heart of the pure, deep connection we seek.

Sometimes, a relationship can begin with utilitarian purpose and change over time to the joy of being together. A business relationship can grow into a deep friendship that continues long past the business purpose. A person who was a possible mate may not have been a romantic fit, but the two people can become lifelong friends. The lady at the checkout counter where you buy milk is perhaps so charming that you return again and again to chat with her over the years, even if other stores are more conveniently located. Shared interests bridge arbitrary divisions that inhibit relationships. Most people think that relationships cannot exist within a one-time interaction or in professional relationships. However, all relationships can be imbued with the joy of being together.

Shared experiences

Shared experiences can fuse transactions into relationships. Experiences with emotional content, shared with another person or group of people, imprint to our memory. Shared experiences can be relived again and again, reinforcing the connection we have with those who shared the experience. Siblings, high school or college classmates, veterans, disaster survivors, mothers of newborns, and numerous other examples illustrate the bonding power of shared experience.

Shared values

Deepening relationships becomes increasingly dependent upon shared values. For example, perhaps you watched your neighbor's son grow up and had fond feelings for him. If asked, you would have endorsed him for a job or a college. Then your daughter one day announces that she and the neighbor kid have fallen in love. Suddenly, you must look at him in a different context, and the need to better understand his values becomes urgent.

Even one-time transactions have a values component. You walk into a store to buy a greeting card. You are in a hurry, but you believe that every person deserves to be seen and respected, especially people in service jobs. The salesclerk across the counter has had a rough day and is at the end of her shift. But she believes her attitude affects her customers and that the last customer of the day deserves her best. The transaction between the two of you begins and ends with a smile, two hearts are lifted, and a small part of the world is a little better for it.

Shared concern (for each other's welfare) or empathy

Empathy is an essential element of rich, long-lasting relationships. As the previous example illustrates, even an empathetic one-time transaction can be memorable and impactful. Empathy is the sensation of walking in another person's shoes. We attempt to understand another person's life situation and feelings. The more we succeed, the more we wish them well and have a desire to help if possible. Individuals vary widely in their capacity for empathy. Those with little or none are limited in the depth of relationships they can form. However, most people can learn to be more empathetic.

Understanding and following the six prerequisites listed in this section can deepen relationships. Introducing them into an existing or new relationship is not difficult. However, kinship is not formulaic, and not all relationships containing the six elements will develop into kinship.

Science to date knows little about exactly what kinship is or how it works. Kinship can only be fully understood when experienced. Many people experience kinship without knowing precisely what the experience represents. Kinship is our ultimate relationship goal, even if we are unaware that it is the goal.

CHAPTER 8

Thinking Like an Entrepreneur

> "Twenty years from now you will be more disappointed
> by the things that you didn't do than by the ones you
> did do. So, throw off the bowlines. Sail away from the
> safe harbor. Catch the trade winds in your sails. Explore.
> Dream. Discover." —Mark Twain

IN "THINKING LIKE AN ENTREPRENEUR," I lean heavily
on a lifetime of entrepreneurial experience and relationships
with remarkable entrepreneurs, and guest lectures and speeches
on entrepreneurship and judging entrepreneurial competitions.
The purpose of this chapter is to emphasize the importance of
an entrepreneurial mindset to your future. An entrepreneurial

mindset will set you apart from the crowd, give you an edge, and is relatively easy to develop. This chapter is directed equally to those already planning or building businesses and to those of you who don't think of yourself as entrepreneurs but will need to think and act like one to maximize success in the future. For the former, it offers some insights you will not likely see elsewhere. For the latter, I hope to change your perception of entrepreneurs and encourage you to develop your own entrepreneurial skills.

For instance, would you have considered Mother Teresa an entrepreneur? Think about it. She broke new ground and set new standards in caring for some of the most vulnerable people in the world. It took great courage and sacrifice, and she inspired millions of people to emulate her work. She, like other innovators, used entrepreneurial skills such as developing a value proposition that inspired her, turning it into a thesis and ultimately a plan, locating financing, building a team, and doing great work.

What Is an Entrepreneur?

A simple definition of an entrepreneur might be: "An inspired value creator," which best describes the entrepreneurs I have known during my long career. This definition is backed up by the fact that investors look for two things above all in entrepreneurs: the soundness of their value proposition and the leadership skills to deliver the value proposition at scale. Entrepreneurs are fueled by an unquenchable desire to create something of value and be rewarded with wealth, respect, changing lives and the world, or all of these.

> Being an entrepreneur is not a career or a business term but a way of life.

The best entrepreneurs are driven to create value that leaves things better than they found them. They realize that failure comes with the territory and is better called *practice*. They get up, dust themselves off, and start over. Overcoming fear of failure is an important quality for an entrepreneur.

Entrepreneurs are critical thinkers. They relentlessly challenge their own ideas and encourage others to do the same. Instead of being discouraged by criticism, they understand that every challenge they overcome strengthens their value proposition and increases their chance of success. Or, if there are too many challenges to overcome, they avoid wasting their time on a faulty value proposition.

There have been entrepreneurs since the beginning of time, imagining and creating something of value for someone else in return for whatever they themselves needed at the time (money, goods, shelter, food, reputation, relationships, etc.). The most important thing being that both parties benefited, and something new was created.

> Innovation and entrepreneurial drive are manifestations of nature's endless capacity to create.

Entrepreneurship is the underappreciated foundation of all commerce and most human activity.

Why the Future Belongs to Entrepreneurs

Global trends—as predicted by futurists, scientists, and leaders in business and commerce—point to a future where an entrepreneurial mindset will be vital whether you are employed, self-employed, or running a business or other organization. Entrepreneurism is the ultimate democratization of work because people can work without needing an employer, and each person has greater control of their own destiny. The entrepreneurial transition is being accelerated by the ability to work from anywhere (WFA) and by the ability to reach markets anywhere in the world through the Cloud with less capital and using virtual companies and gig-work staffing. As you live longer and have more jobs and careers during your lifetime, you will almost assuredly either become an entrepreneur or need the mindset and skills of one. As explained in part one of this book, democratization causes disruption and deinstitutionalization of industries and organizations, and disruption is the business of entrepreneurs.

Entrepreneurism is a worthy and beneficial calling that fuels commerce and economic expansion and improves the lives of millions. Entrepreneurial opportunity can result in historic worldwide prosperity. In India, for instance, members of the Scheduled Castes (previously known as Dalits, Harijans, and Untouchables) are witnessing a phenomenal rise in entrepreneurship. In the western part of Uttar Pradesh, India's most populated state, entrepreneurship by Dalits has risen by almost 300 percent in the last twenty years.*

As more people worldwide rise above subsistence living, they form a growing wave of eager new consumers and entrepreneurs.

---

* Saurabh Rai, "Breaking Caste Barriers: Stories of 5 Dalit Entrepreneurs Who Reached the Top," SocialStory, January 25, 2018, https://yourstory. com/2018/01/dalit-entrepreneurs.

A Cloud-based digital economy expands the opportunity to self-educate and start a business at little to no cost anywhere in the world. For investors, bankable value propositions and leadership skills supersede education, ethnicity, and even geography in the future. As Elon Musk said recently, what matters most in employees is a pattern of performance. Many people, especially in the West, fail to appreciate that it is entrepreneur-driven innovation that is most responsible for improving the world and people's lives rather than government programs or even well-meaning non-governmental organizations, although they play important roles. Most people who fail to understand entrepreneurship and innovation's central role in lifting humanity have rarely created a product or business from whole cloth, been an entrepreneur, been responsible for a payroll, or led a business. That group, unfortunately, includes many politicians, regulators, government workers, and other policymakers.

The United States is the best place to be an entrepreneur

The United States for decades has been the best country in the world to be an entrepreneur; we are a country of pioneers, explorers, and entrepreneurs. The United States offers the world's most welcoming infrastructure for entrepreneurs in terms of research and development, financial innovation, and potential for reward, and it's only going to get better due to technological advances, human empowerment, and demographic change. That entrepreneurial spirit, opportunity, and infrastructure is the golden goose that has made the United States the recipient of more hopeful immigrants than anywhere in the world.

Innovation itself is being democratized

One of the best examples of the democratization of innovation is open-source technology, where source code is made available freely for people to use and distribute publicly and collaboratively. Innovative concepts are increasingly being widely shared, allowing people access to the tools they need to be successful.

This is in stark contrast to our industrial past, which was dominated by patents, closed and often crony-based supply chains, and monopolies. All those are still very much a factor in today's business. Many twenty-first century entrepreneurs have a different philosophy, however. They believe that while having a successful product or company is good, empowering an entirely new industry that validates their company or product is better. If you are a smartphone manufacturer, for example, promoting open-source technology and smartphone apps is like having an R&D department with millions of innovators and a collective budget that you could never match to improve your product's value proposition. Tesla and Elon Musk seem to believe that if the electric vehicle industry prospers, they will as well and tend to embrace rather than fear or retard competition.

The risks between employment and an entrepreneurial career have equalized

More people are choosing entrepreneurial careers over employment, and the trend will continue. In the past, job security meant working for a solid company for as long as you could, ideally for life. All that has changed; employment is no longer the golden ticket it once was. Billionaire founder of Paychex and entrepreneur Tom Golisano states in his *Wall Street Journal* bestselling book, *Built Not Born,* that there is no longer any material difference in the risks associated with employment and an entrepreneurial career. Why is that the case? One of the primary reasons, as discussed in Chapter 2, is that companies and institutions today are at increased risk of disruption and deinstitutionalization. Employment security is a thing of the past; companies will become increasingly less stable and will appear and disappear faster than at any time in history.

Of course, not all companies will falter; some large corporations and conglomerates will prosper, but the attraction of long-term job security and career growth within a single firm will no longer be the draw it once was. The "gig" economy will become increasingly popular and desirable. Those with marketable skills will choose to work for multiple customers, rather than being dependent on a single employer, a single executive decision, or the threat of a company's products or services becoming obsolete in a fast-changing world.

Case study: Luke's story

In early 2020, Luke was in his mid-thirties and had a master's degree. He was employed as a securities research analyst. Unexpectedly, he received a call from an old friend working at a university that was involved in a multiyear securities research project. He invited Luke to join the new team. Luke had long considered getting his doctorate. He was able to negotiate pursuing his doctorate and flexible work hours as a condition of the new job. Luke recognized that the world was changing rapidly and an opportunity to be on the leading edge of new research for the next four years would build his resume, allow him to learn new skills, and get specialized experience—all good for his career.

When Luke gave his employer notice, they attempted to retain him with an increased salary, but Luke was resolute that he wanted something new and more control over his life instead of the long hours his current employer required. Thinking on his feet, however, Luke proposed to continue to assist his current employer as an independent contractor part-time.

The company accepted his proposal. Under the new arrangement, Luke's income was adequate for far fewer work hours. The new university job was not nearly as demanding, and he was able to manage his new full-time university job, part-time gig work for his old employer, and pursuing his doctorate.

Benefits may become portable

The perceived risk of going out on one's own was one reason people favored employment in the past. However, employee benefits were possibly an even greater factor.

US law in the form of the Employee Retirement Income Security Act of 1974 (ERISA) favors employment over self-employment and individual business ownership and large businesses over small enterprises. This stifles economic growth and entrepreneurship in ways never intended when the law was passed. Government policies like ERISA are an example of outdated, industrial-age laws and regulations. Forward-thinking politicians should be advocating changes to ERISA that support the reality and economic opportunity of a digital economy with many more entrepreneurs and gig workers.

In an entrepreneurial economy, employee benefits should be based on individual (personal) ownership, and they should be portable. To some degree, this has been achieved with pensions through 401(k) plans and individual retirement accounts (IRAs). However, healthcare benefits—which are the most costly and important employee benefit—are woefully lagging. The United States and other countries need a health plan system that encourages, rather than discourages, entrepreneurship, such as one modeled after the 401(k) plan rather than tying benefit availability and cost to an employer, particularly favoring large employers over small employers or self-employed.

An economy that is entrepreneurially oriented will be more stable and innovative. Governments at all levels worldwide would serve their citizens well by abandoning industrial-age practices, regulations, and laws to favor the age of the enlightened entrepreneur.

Entrepreneurial financing has never been more readily available

I can imagine at this point you are thinking, *All this is well and good, but what about financing?* Financing for entrepreneurial activity has dramatically expanded in the last few decades and will continue to expand as the global population ages, more people save for retirement, and a portion of their savings are directed to funding entrepreneurial activity. Unknown to most people, pension fund investment managers typically invest 5 to 10 percent or more of retirement funds in "alternative investments" that are predominantly entrepreneurial businesses because they offer higher long-term returns.

If more people understood that a substantial portion of people's wealth is directed to entrepreneurial companies that generate higher investment returns, new jobs, and fuel the economy, politicians who support capital gains taxation would receive fewer votes. Capital gains taxation reduces retirement income while hampering the very activity that fuels innovation and promotes a healthy economy.

Types of Entrepreneurs

The term *entrepreneur* has many interpretations and connotations, but it is usually associated with business founders. Most people envision a Steve Jobs or Bill Gates starting a business in their garages, or a teenage Elon Musk launching a rocket. However, entrepreneurs come in all shapes and sizes. Consider, for example, independent realtors, insurance agents, micro-enterprise business owners, the mom-and-pop store on the corner of your street, or corporate leaders who have grown small companies into major public companies.

There are also entrepreneurial employees who are financially or emotionally invested in their employer, people who don't just show up and blindly work their eight-hour shift and go home. They understand that their actions can make their organization or company more successful and behave as if they are owners. These individuals are already sought after by employers, but they will become increasingly valuable in the future.

Lifestyle versus wealth building—your choice

If you are considering a career as an entrepreneur, your first decision should be whether your goal is to make a good living or generate significant wealth. There is nothing wrong with either choice, but it is important to have a clear financial goal from the outset. *Lifestyle entrepreneurs* build businesses to support their lifestyle or generate a second income in addition to their "day job." On the other hand, *wealth-creating entrepreneurs* look for value propositions with high-growth potential, resulting in a company that they can sell to a strategic buyer or that they can convert to a publicly-traded stock company.

There are substantial differences in the interests, skills, experience, and drive of lifestyle entrepreneurs compared to a wealth-building entrepreneur. The demands placed on each are also substantially different, as is the level of personal risk they take.

An example of these two different types of entrepreneurs can be illustrated in the stories of two successful brothers with different objectives. One was an honors graduate at Purdue University who landed a well-paying technology job after college. After fifteen years as a successful technologist, he realized that becoming a manager was the only route to earn more money and gain more control over his future. It didn't take him long to realize that he did not enjoy managing people and did not feel in any greater control of his future. He decided to build his own business after considerable soul-searching about his skills and the work he enjoyed. Research over several years led him to choose becoming a financial advisor despite the industry being highly competitive. Today, he is highly successful, respected in his field, and thrives on the trust his clients place in him and the measurable value he delivers.

The other brother discovered, after competing as a high-school and college quarterback, that he had leadership skills, an ability to keep cool under fire, and that he enjoyed having accountability to his team and the fans. He began to envision a future as an entrepreneurial business leader during his college years. Recognizing that he needed to learn the craft of business first, he spent two decades in sales and corporate leadership roles before launching his first entrepreneurial venture. Today, he is the respected entrepreneur/CEO of his third successful business. Here we have two guys from the same gene pool who are both successful entrepreneurs loving what they do, but with different endgames that fit their personalities and life goals. Let's explore six types of entrepreneurs.

Sideline or hobby business

Many entrepreneurs earn extra income and diversify their income sources by holding down a full-time job while using their innate talents to pursue their entrepreneurial passions or dreams. Sam is a fireman. His wife, Theresa, works as a hairdresser. Both were in their mid-thirties when this story began. They treasured their time together away from work because their work involved odd hours that were challenging to coordinate. They had lots of energy and were ambitious, and they were determined to create the financial security for a better life. Both were talented with their hands. In 2016 they were able to buy a tiny, rundown duplex near their home. They spent their hours together refurbishing and remodeling the duplex into a rental property. Things went well, and success with the first property led to buying another property two years later and a third in 2021. They used the rental income to pay off the earlier mortgages and set aside cash to invest in other properties. They did not spend any of their real estate income to improve their current lifestyle; they put it all on the future. The real estate boom of 2020 increased the value of their owned properties, which enabled them to get a mortgage to buy a fixer-upper that could be attractive enough for a vacation rental if they put enough work into it. They succeeded. Using their social media skills, they did their own marketing to keep the vacation rental occupied and keep their operating costs low. Sam and Theresa have an ambitious but realistic goal to own $5 million in real estate free and clear by the time they retire.

Not all entrepreneurs are adults. In July 2020, Scottsdale, Arizona, native and elementary school student Nicholas Bubeck founded a company eponymously named Creations by Nicholas. The seven-year-old created arts and crafts kits for children featuring model airplanes made from popsicle sticks, corks, bottle caps, and other household items. In his first year, he sold almost one thousand kits, donating a dollar from each to charity, demonstrating that Nicholas was not only a natural entrepreneur, but an enlightened one as well. Nicholas is already on his way to reaching his potential, thanks to his mom appreciating that entrepreneurial experience would be invaluable for a lifetime.

In Canada, ten-year-old Domenic LaHaye from Saskatchewan started a company making plant potholders from reclaimed grain elevator wood. His love of plants began when his teacher brought several plants from home to brighten up her students' classroom. His mother supported his desire, thereby launching another young entrepreneur. La Haye's company, Little Holders, has a logo, business cards, and a great future.

Novice entrepreneur or independent contractor

There are several classifications of business entrepreneurs, and each inhabits a slightly different spot in the business world. First, we have the independent contractor—think plumber, drywaller, electrician, or consultant who becomes self-employed. They probably worked for a company and decided at some point to break out on their own. Some remain single-person businesses; others grow and begin hiring staff. These entrepreneurs usually have several things in common: they have one company or legal entity, they are small to midsize, they operate in one country only, in a single industry, and their business financing is straightforward.

Thomas is a good example of this type of entrepreneur. He worked in a series of auto dealerships for twenty years in various service roles, including management. He loved the work and was well paid, but the auto dealership industry had been rapidly contracting due to consolidation and democratizing innovations such as Phoenix-based Carvana and Tesla who sell cars online or through showrooms that bypass dealers. After repeated dealership closings and cutbacks, and more and more hours demanded without pay increases, Thomas was burned out, bored, and unfulfilled. He wanted to enjoy life more with greater job security. Thomas heard that one of his wealthy automobile customers was looking for a butler-cum-groundskeeper-cum-property manager. That was several years ago. Today, Thomas has several clients and manages a slate of upscale properties with his brother. The most important thing? He loves his job, feels fulfilled and more secure, and has far more freedom to pursue other interests and hobbies.

Career entrepreneur

Another type of entrepreneur is the person who inherits or launches a single company and grows it over several years (a one-time or one-off business). The businesses can be large, or very large, and they can be international, but they are limited to a single industry and were launched using traditional financing. Many of these companies are extremely successful and make huge profits because they are highly focused and skilled in a single niche.

Experienced (or serial) entrepreneur

These entrepreneurs are innovators who repeatedly identify value-creation opportunities and have the managerial skill to turn them into successful businesses. Some are small to midsize, and others grow to become industry leaders. They may stick to a single country or expand internationally. Serial entrepreneurs know how to use a variety of financing that best fits each entrepreneurial venture.

Master entrepreneur

Finally, we have the *master entrepreneurs*. These are folks of rare talent who have a track record of creating successful companies in a diverse range of industries using the most effective and often complex financing approaches. Some operate in a single country and others internationally. There are many well-known master entrepreneurs, but two of the most famous and successful are Elon Musk and Richard Branson. Musk deserves a special mention because he is what one news commentator called "a national treasure" and who author Peter Diamandis referred to as "the greatest entrepreneur of our age." Musk is an inventor/innovator well-grounded in science who has broken new ground and done what others thought impossible in three industries thus far. He is also an enlightened entrepreneur who passionately wants to leave the world a better place, beginning with establishing a stimulating and rewarding work environment for employees. Branson is also a wunderkind who has created a distinct brand, along with a powerful business strategy and operating style, and he is equally committed to leaving the world a better place. There are other master entrepreneurs who are not as famous but are no less talented. I have been privileged to know many, and one, Mickey Maurer of Indianapolis, is among my best friends. I'm a good entrepreneur but not a master, and I watch in awe as these folks do their thing.

You, as a Product or Service

Everyone has ideas, innate talents, and passions as the previous case studies illustrated. An entrepreneur is never satisfied simply to have a good idea. They are driven to pursue those ideas and passions by turning them into products or services that people value. They are unsated unless that happens. In essence, we are all acts of creation, so creativity is part of our core being. Creating something is incredibly rewarding, a positive capability hardwired into every human being to be discovered. You can see the joy of creation in the eyes of a child as they draw their first tree or dinosaur. That joy is an innate response, not learned. Never let anyone tell you that you are not creative. Do not worry about being unconventional. To be unconventional can be an expression of our aspirational drive to grow and reach our potential. Leaders and innovators are usually unconventional.

In terms of thinking like an entrepreneur, we each have our own "zone of genius" from which ideas can spring; any one of which may be turned into a commercial value proposition. When that happens, we have something that we can sell. All the entrepreneurs you've been introduced to here and all those that I have known in my career were everyday people who tapped into and trusted their "zone of genius." Perhaps you doubt that you have the energy and drive to be entrepreneurial. Maybe it is because you never had a cause, an idea, or a business that was entirely yours. Entrepreneurs sometimes begin as bored, unchallenged, underachieving students or employees unmotivated by what they are asked to learn or the work that they are asked to do. Trust that you will find bottomless energy and motivation once you unleash your creative core and own the results of your labor.

In the twenty-first century, we need to think about our zone of genius as a source of unique commercial value. It is the opposite of being an indistinguishable, interchangeable spare part in an industrial-age corporate machine. You can either lease your zone of genius to a company as an employee, or you can package it into products and services that you sell to others. Trust your instincts and entrepreneurial mindset to launch your zone of genius into the future and toward your potential.

PART III

Skills and Mindset for Life in the New Reality

THE PREVIOUS FOUR CHAPTERS ADDRESSED important considerations with a focus on New Reality careers. Preparing to live in the New Reality requires equal attention.

Part III addresses three essential twenty-first century life requirements: health, wealth, and success. These have always been important, but the New Reality creates unprecedented urgency that requires a mindset change and new practices. Paying serious, deliberate attention to your health, wealth, and a new definition of success will be essential. Failure to do so will have unpleasant consequences. The recommendations are to prioritize and optimize each to a degree never achieved by prior generations. Your ancestors valued health, wealth, and success, but most of them never gave them the priority and attention they warranted. By neglecting them, they experienced regrets, disappointments, and needless suffering later in life that could have been easily avoided if addressed earlier.

As I wrote this book, it was painful to reflect on the needless suffering that I have observed in life because people failed to develop healthy habits younger in life and paid dearly in their last quarter. Far too many people struggle financially, retire in poverty, and spend their final years in a poor-quality nursing home, financially destitute because they did not learn when young to save and invest instead of spending everything they make. Far too many people are physically older than their years, miss out on enjoyable activities with friends and family, and develop lifestyle-related chronic diseases that fill the last quarter of their life with worry, pain, and suffering—all because they did not care for their bodies early in life with proper nutrition, exercise, sleep, and stress management. Far too many people devote their life to unfulfilling careers and jobs for financial and social status that disappoint them—all because they defined success in a way that would never satisfy them. The next three chapters offer suggestions for avoiding these all-too-common, unnecessary miseries.

The New Reality offers you the unprecedented opportunity for the highest quality, most meaningful life in history. You can maximize that opportunity by pursuing health, wealth, and success with practical methods you can easily master. You have likely already achieved things far more challenging. The rewards for your effort are an unprecedented opportunity for freedom, joy, and wisdom. I hope you will find these next three chapters relevant and be motivated to pursue the bounty residing in your potential.

CHAPTER 9
Health

["The first wealth is health." —Ralph Waldo Emerson

LIFESTYLES BECAME LESS HEALTHY DURING the twentieth century as work and peoples' lives became less physical, more sedentary, and increasingly stressful as the pace of life and change accelerated. During the same period, food became less expensive and more plentiful but less nutritious.

Concerns have been growing for a half-century about unhealthy lifestyles leading to long-term, incurable, multi-chronic diseases and conditions in the elderly. The concern is accelerating because of the unsustainable demands that a growing, aging, unhealthy population places on the healthcare system, families, and national economies. The World Health Organization (WHO), the US Centers for Disease Control (CDC), academic medical centers such as Harvard and Stanford, and esteemed healthcare organizations such as the Mayo Clinic and Cleveland Clinic report disturbing health trends in the global population. They identify unhealthy lifestyles as the leading cause of chronic diseases and conditions in the elderly that could lead to a financial and healthcare resource crisis in a worldwide aging society.

However, it is a crisis that can be averted. Harvard Medical School recently published the results of a study of 120,000 people from the 1980s to 2014.* The study analyzed the impact on longevity (after age fifty) of five lifestyle choices: healthy diet, moderate exercise, healthy weight as measured by body mass index (BMI), moderate alcohol usage, and never having smoked. People in the study group conforming to all five healthy lifestyle indicators at age fifty lived longer than those with none of the indicators present. On average, men lived twelve years longer and women fourteen years longer. Equally impressive, each indicator added two years on average to a person's longevity even when others were not present.

The Coming Change

How can we define what a healthy lifestyle looks like? What is ideal?

> *Optimal health* is a commitment to maximizing physical, emotional, spiritual, and relational well-being, the opposite of an unhealthy lifestyle and mindset.

---

* Monique Tello, MD, MPH, "Healthy Lifestyle: 5 Keys to a Longer Life," Harvard Health Publishing: Harvard Medical School, March 25, 2020, https://www.health.harvard.edu/blog/healthy-lifestyle-5-keys-to-a-longer-life-2018070514186.

People do not tend to make behavior changes, even important ones such as living healthier, when they perceive the benefit to be far into the future. They need to see an immediate result. However, their attitude changes when near-term consequences become unacceptable and economically penalizing or infeasible. And more importantly, their attitudes change when they become convinced that optimal health is achievable without extreme effort or inconvenience.

Living and working longer in better health

> Most people under age fifty today will see their one-hundredth birthday. Those under age twenty-five will possibly live a quarter of a century longer.

Longer lives proportionately extend the stages of life and alter behavioral and cultural norms. For instance, even today, young adults marry and start families later. The Pew Research Center reported that 52 percent of young adults aged eighteen to twenty-nine live with their parents, the highest rate since the Great Depression. Many middle-aged and elderly people today are more active, healthier, and retire later—if at all. This is a positive trend because as people live longer, they will need to work longer and need optimal health to do so. The traditional eligibility age of sixty-five for social security, Medicare health coverage, and employer retirement plans in the United States will almost certainly be extended because it will become fiscally impossible to retain age sixty-five eligibility when there are not enough younger workers to pay taxes and subsidize the insurance premiums of a growing elderly population.

In addition to the economic reasons to work longer, few people approaching retirement age today find a long traditional retirement appealing. Instead of retirement, they seek "refinement," their best years instead of their worst, expressed through greater freedom, personal growth, and significance. Refinement will become increasingly possible earlier in life through work-from-anywhere (WFA) careers and lifestyles and gig work. Demand for older adults as entrepreneurs, gig workers, and employees will increase as a result of the declining younger workforce and the creativity, maturity, and wisdom that older workers bring to the New Work.

People of all ages are beginning to look at health and healthcare differently than people in the past, with a perspective best described as "affordable healthy longevity." This could be defined as living a long, healthy life followed by a short period of frailty and dependency before death. Commitment to optimal health today can make affordable healthy longevity possible for many people.

However, as a society, we have a long way to go. Healthcare today is not making us healthier. The system works to keep people alive longer while simply managing their chronic health conditions at great expense to them individually and to all taxpayers. Approximately 5 percent of the US population with chronic and complex diseases and conditions are responsible for roughly 50 percent of all healthcare costs.

Healthcare challenges in an aging society

The aging population healthcare crisis will peak sometime between 2045 and 2055 as baby boomers reach their frailer elderly years; a crisis will slowly emerge of too many frail elderly people and too few young people to pay the taxes and insurance premiums that subsidize their healthcare costs.

Medical innovation and automation are the best hope for relief, but there are no breakthroughs on the horizon that are of sufficient consequence to avert the crisis. The general public is unaware of the healthcare crisis that is about to descend, how it will affect them, and what they can do to mitigate its effect. Perhaps something will emerge, but as responsible individuals, we cannot count on it.

The consequences for lack of knowledge or inaction can be severe. If you live through the next thirty years or so, it may become increasingly difficult to gain timely and affordable access to physicians. Delays in treatments are likely to increase as demand for healthcare services grows. Self-care and homecare (using remote care technologies) will become progressively necessary. Health and healthcare will consume an increasing share of your disposable income. More than ever before, individuals and families will need to create and manage their own financial healthcare reserve for medical costs not covered by government or private insurance. Financial reserves will also be required for medical professionals to provide advocacy and coordination of care for complex diseases that cross multiple specialties, for lifesaving or life-extending treatments not covered by government or private insurance, and for elderly residential living.

Timing Is Ideal for Optimal Health

Timing is ideal to adopt an optimal health lifestyle. Healthy lifestyles are increasing. I have never witnessed anything like it in my forty-plus-year career in the healthy living and healthcare industries. Cultural encouragement has played an important role in the change because we are motivated when we see others living healthy lifestyles. Social and cultural indicators point to healthy lifestyles becoming a cultural norm over the coming decades.

Services and innovations to optimize healthy living are expanding. More and better exercise programs, health monitoring, nutrition, stress relief, and sleep aid innovations are available every day. The COVID-19 pandemic unleashed the power of mobile, remote healthcare and expanded its use in ways that would have otherwise taken decades. It also demonstrated to millions of people that greater self-sufficiency was possible during periods when care from hospitals and doctors was unavailable or difficult to access.

Technology providers are responding aggressively to the post-pandemic opportunity. Wearable and implantable sensors will create personalized health baselines, monitor health in real-time, and even offer treatment options. Apple Watch sales, for example, are increasingly driven by the health benefits it offers. Apple appears committed to its watch becoming an essential optimal health device.* As the Internet of Things (IoT) and smart homes become a reality, temperature checks, urine and stool samples, and weight and blood tests that today require a doctor or lab visit will be provided automatically by our smart homes. For example, smart home bathroom mirrors will include technology to check temperature and facial changes that signal extreme stress or stroke risk. Bathroom toilets can perform urine and stool sample tests. Smart home monitoring results will be fed to artificial intelligence systems that look for aberrations in our health baseline and notify us and our physician if they appear. All these innovations support healthy longevity that will enable people to live and work longer, as demonstrated in Peter's story in the prologue.

---

* Rolfe Winkler, "Apple Struggles in Push to Make Healthcare Its Greatest Legacy," *Wall Street Journal*, June 16, 2021, https://www.wsj.com/articles/apple-struggles-in-push-to-make-healthcare-greatest-legacy-11623832200.

Health Mindset

It may seem counterintuitive that optimal health begins with our minds instead of our bodies, but health improvement attempts are rarely sustainable otherwise. An unhealthy mindset is the norm today; it became culturally embedded during the twentieth century and continues in many people today. It must be set aside if we are to face the New Realities of the twenty-first century.

The "it's too difficult" mind game

Perhaps the single greatest barrier to optimal health is fearing that it requires a lifetime of denial and suffering, something which most people find unattractive and unsustainable. A more apt metaphor is climbing a steep hill; it is tough at first, and we may want to give up, but as we struggle on, increasingly determined, the climb becomes easier. Upon reaching the summit, we enjoy the magnificent view and walk down laughing and reliving the victory.

There is a moment when you are moving toward an optimal lifestyle when everything changes; typically, this occurs anywhere from two to twelve months after beginning your journey. Suddenly you are aware that even on a lousy day, where everything is going wrong, you feel better because you made the effort to live healthier. The reverse is also true. The days that you neglect your health feel incomplete or dissatisfying, even if it was an otherwise successful day.

The procrastination mind game

This mind game allows us to accept that we *should* live a healthy lifestyle, and, in fact, plan to . . . *someday*, just not now. Our excuses for inaction often center around lack of time or money. It is most often neither. Instead, health has been assigned a lower priority than time or money. If you think about it, time is irreplaceable but has decreased value if our health is poor. Health has an unlimited ability to produce wealth, but wealth has limited ability to produce health. Delay in optimizing health can have severe consequences. We see this when people wait too long to obtain a medical diagnosis. If they had faced the issue earlier, the consequences could have been minimized.

The avoidance-rationalization mind game

In this mind game, we accept that taking care of our health is absolutely necessary—for everyone else. We, on the other hand, are too far gone—too out of shape, overweight, or in too poor health to be improved. Or we may be fatalistic and rationalize we are going to die anyway, so why worry about our health? Life is short, so why not enjoy it while we can? To paraphrase deceased comedian George Carlin, wouldn't it be awful to live until you are one hundred years old and then die of nothing? The problem with this thinking is that you are more likely to live decades suffering from multiple painful chronic diseases than to have a quick and easy death. More importantly, such thinking ignores the present-day benefits of healthy living.

One of my mentors wisely said that every morning when we awaken, we make the most important decision of our lives: we choose to grow or to begin dying. Healthy growth rewards us with chemical messengers (endorphins) that make us feel uplifted. Refusal to grow diminishes self-respect. The reality is that health can be optimized for almost anyone regardless of age, health status, financial condition, or available time. Anyone who thinks otherwise should attend a Special Olympics event, watch a disabled athlete in action, or observe a SilverSneakers class* at a local fitness center.

The mind game of mimicking other people's looks

In this mind game, our motivation for becoming healthy is based on trying to look like people we admire, rather than to feel better today. Younger adults are particularly susceptible to this mind-game. Mimicking movie stars, supermodels, or athletes is rarely a sustainable motivation because we are not, and never will be, them. Striving for optimal health means playing the cards we have been dealt and challenging ourselves to achieve our unique potential. Everyone's baseline is unique as are their health challenges. A three-block walk to the store may be a supreme effort for one person and be equivalent to a ten-mile run for an athlete.

The impatience mind game

In this mind game, we accept that we should change our lifestyle, but then place impossible conditions upon success that guarantee we will soon give up. For example, attempting to go from never exercising to exercising two hours per day or attempting to lose thirty pounds in two weeks. It takes years of unhealthy habits to create the bodies and mindset we have; we are not going to change them overnight. So, how do you begin the process? Start making small healthy decisions and setting incremental goals—small wins to create a success mindset. Every day provides opportunities to make healthy choices that should be seized no matter how small. For example, park farther away from store entrances to walk more.

---

* SilverSneakers is one of the world's most successful fitness programs designed for seniors seventy-two and older, many of whom had no prior fitness experiences.

Take fifteen minutes to walk, stretch, do a yoga pose or two, or do push-ups when you don't have time for more. It is the incremental progress we make that builds the confidence we need and encourages us to do more.

A Simple Plan for Optimal Health

Optimizing health is based on integrating the following five components. You can build a plan designed just for you that matches your personality, baseline body, and lifestyle using these components.

Total or holistic health

Our physical, mental, and spiritual health are interrelated. You may feel physically and emotionally healthy but still long for meaning. You may suffer emotionally and feel loss of meaning but be comforted by your level of fitness. When all these elements work together, you will be inspired to do more.

Our physical body needs to be appreciated as a miraculously engineered, adaptive, resilient whole that needs to be fed, watered, exercised, rested, and loved as you would a child or a pet. Abused or neglected, it will strike back with pain, illness, or death. It is an abuse of nature's gift to neglect your body, and you should abhor this neglect as you would that of a child or pet.

The psychological body is an exquisite organism like the physical body. By the time we are twelve years old, we have learned about our physical body from biology classes, sports, and family doctors. However, we can reach middle age and know little to nothing about our psychological body or how to optimize its functions and systems. The psychological body needs equal respect and attention if we are to lead a healthy life, and it is easier to understand than you might expect. The second book of The Potentialist series addresses the psychological "body."

Spiritual health is the innate, healthy need to answer the unanswerable and fathom the unfathomable. We discover awe and respect for life in the process of such contemplation and are comforted and strengthened afterward. Spiritual health and religion can be complementary but are not synonymous. It is possible to be spiritually engaged without being religious and vice versa.

The four foundations

A healthy lifestyle requires attention to four foundations: physical activity or exercise, nutrition, sleep, and quiet time. We tend to think of them separately, but they are highly interrelated. Exercise is less effective when we do not sleep well or are over-stressed. We sleep better after exercise, a nutritious meal, and quiet time. Form matters also. An hour of exercise done incorrectly is less effective than fifteen minutes in correct form. When and how we eat matters as much as what we eat.

Measurements

Most businesspeople subscribe to the axiom, "If it isn't measured, it doesn't exist." Many people who practice that rule in their professional lives resist spending three to five minutes per day to track their exercise, nutrition, quiet time, and sleep. Measurements (keeping track) reinforce living to optimal health. Daily and long-term goals should be achievable, but they should be a stretch. For example, a weight goal could be attached to a desired weight at graduation or marriage. An endurance goal might be a repeat of a tough hike on the anniversary of the first one. Today's smartphones and smart watches especially make keeping records and measurements easy.

Customization

Don't be timid about customizing and creating your own plan. The more you own it, the better you will follow it. The best exercise is the one you will do. The best diet is the one you will follow. The best way to find quiet time and relieve stress is the method that most often works for you. Sleep will come when you help it out with routine and preparation. Calmness and steadfastness win over extremes: everything is better in moderation. Perfection is the enemy of good; be patient with yourself but keep records so you don't kid yourself about progress.

Variety and consistency

Humans crave a balance of variety and consistency. Variety prevents boredom, and routine builds strength and endurance. Our hyper-efficient bodies learn to malinger unless we cross-train when exercising. Too much exercise repetition leads to injury, so exercise programs should rotate strength building, high-intensity aerobics, low-intensity aerobics, and flexibility. Our bodies crave a balanced variety of foods and colors; diets that fail to offer them become unappealing. Daily meditation is an enriching habit, but even more so during or after a nature walk.

Need a draft baseline?

The lifestyle plan that I have followed since 1999, along with the measurement forms and reports that I use, can be found at www.potentialistfuture.com. My plan is loosely based on the books *Body for Life* and *Eating for Life* by Bill Phillips, and the work of several other authors and healthy-living experts, customized with my own research and experiences as a lifelong athlete and as co-founder of a highly successful prevention and wellness business. I am confident that I would not feel as good and have my current level of energy and mental acuity had I not created my own optimal health program in my late thirties. Many of my then-healthy contemporaries did not and are either deceased or are in poor health today.

Records matter. Goals matter. The discipline to eat and drink well, but modestly, matter. In my mid-seventies, I am four pounds heavier than when I graduated high school. I lift as much weight as I was lifting in my mid-forties. I still am up for hikes with my young adult grandchildren and can press them to keep up with me on bicycles. This reality isn't an accident, and there are millions of people my age and older far more fit than me.

> Optimum health is achievable regardless of your current health status.

Please feel free to use the contents on the website as a baseline for your own fitness if you find it useful, but always build a plan that is your own. That's essential.

If you put this book down at the end and do nothing else, please consider prioritizing an optimum health lifestyle. Health is the wellspring from which all else in life emanates.

A Tale of Courage, Determination, and Incremental Success

A few years ago, I witnessed an unforgettable act of courage from a mountaintop bar-café in Italy's Dolomite Mountains. The bar was equipped with telescopes, so along with about two hundred other people, I watched a women's biking group snake up the mountain from the valley below. Most of the bikers appeared to be in their twenties or thirties. One woman (whose name I later learned was Ruth) was at least twenty years older than the other women and was obviously not an athlete like most of them. As the ride up progressed, Ruth continued to fall behind. She would periodically dismount and push her bike up the hill for a few minutes and then mount up again and continue pedaling. Those of us watching Ruth from the bar-café became mesmerized by her determination and spontaneously began to cheer her on. Some of the people in the café went down the mountain to walk alongside and provide encouragement. When she reached the summit, Ruth received a tumultuous welcome of applause, backslaps, and hugs. She was so overwhelmed, she climbed atop a picnic table to thank everyone. She said the ride was motivated by the recent and unnecessary death of her younger sister due to complications from obesity. She was, she said, determined to avoid the same fate. She commented on the many stops she made as she climbed the mountain road, explaining that when she felt she could go no further she stopped for a short period and then recommitted to just fifteen minutes more. These multiple small commitments led her to success. We later learned Ruth's age—fifty-eight. There is a lot we can learn from her. I think two hundred people learned something that day. I know I did. There is a Ruth inside all of us; let her be your inspiration for optimal health.

CHAPTER 10
Wealth

> "If we command our wealth, we shall be rich and free; if our wealth commands us, we are poor indeed."
> —Edmund Burke

THE NEW REALITIES DISCUSSED THROUGHOUT this book create unprecedented challenges and opportunities to everyone's personal financial situation, including ten to twenty or more additional years of life; multiple jobs and careers over a longer work life; periods without income; and a need for larger financial reserves for contingencies and opportunities.

Insufficient savings, contingency reserves, and liquidity combined with excessive debt are all too common and lack the flexibility needed for the times ahead. Even astute financial managers will be hard-pressed to adapt, as the institutions and practices that have guided the world economy for decades are themselves redefined.

Financial planning for the times ahead need not be uncertain and frightening.

> Almost anyone can build the financial security for a lifetime if they start early in life; set clear, achievable goals; and maintain spending discipline.

Yet far too few people achieve wealth and security today. The typical assumption is that they do not earn enough money to build wealth and become financially secure, but millions of low-income families do it while millions with higher incomes fail. The real problem is that most people lack a wealth mindset. They live for today, spend beyond their means, and ignore the consequences of stress and loss of peace of mind. As a result of living too much for today and too little for tomorrow, instead of a healthy balance, they are unmotivated to plan and invest early in life to build wealth and security effortlessly. This chapter examines why, outlines a New Reality financial mindset, explains the fundamentals of investing and selecting an advisory team, and suggests a new life goal called refinement to replace retirement. The contents of this chapter represent my opinions and experience. The chapter is not offering investment advice, nor does anything in this chapter replace advice from a professional financial advisor.

Wealth Mindset

> The first step in developing wealth is to understand that your current mindset may be holding you back.

Many people are unhappy with their lives despite historically high standards of living. The benefits of wealth are attractive, but many people believe it to be beyond their reach and do not act to build it. Additionally, people with wealth can experience unexpected problems that offset much of its benefits. Mismanagement of wealth causes needless losses and missed opportunities. The examples in this section are intended to help you assess your own mindset about wealth.

Equating wealth with what it cannot provide

Despite overwhelming evidence to the contrary, many people equate wealth to success and happiness. Culturally entrenched mindsets and practices border on liturgy. Psychologists, researchers, and authors have explored this millennia-old distortion. For example, author Lynne Twist in *The Soul of Money* describes "the toxic myth of scarcity," in which people become trapped in a mindset of never having enough, whether that is money, time, sleep, love, or anything else. They come to treasure what is scarce, such as jewelry, expensive homes, clothes, and cars, while ignoring or taking for granted the abundance right in front of them through relationships, health, nature, and the highest quality of life in history. Whatever they have, they always want more in an endless meaningless cycle of chasing what they do not have, capturing it, devaluing it, and beginning the chase anew for something else.

Twist also discusses the destructive confusion between poverty, wealth, and happiness. She is well-qualified as a founder, fundraiser, and field worker at the Hunger Project, serving the world's most economically disadvantaged. She believes that no one should be called "poor" because richness in spirit is frequently found in the economically destitute, but often missing in the wealthy. Mother Teresa made similar observations. People who saw the virtual reality film of the Indonesian homeless woman living in a dumpster discussed in Chapter 7 reported her to be remarkably happy and positive. As a kid on a Texas ranch in the 1950s, many of my neighbors were subsistence-level ranchers and farmers, but they were proud, happy, and never considered themselves poor. They took pride in doing their best with what they had, in hard work, in their families, neighbors, religion and communities, and in working in nature.

Secrecy

Today, people discuss sex more openly than money. In many cases, people are uncertain of their financial competence and are hesitant to admit it. Reluctance to discuss personal financial details is understandable and at times prudent, but discussing principles and practices with knowledgeable advisors, friends, and associates is a valuable learning opportunity. Secrecy and unwillingness to ask for advice, or seek professional help, is a frequent cause of avoidable financial loss.

The grasshopper and the ant

One of Aesop's fables tells the story of a starving grasshopper who asks for food from an ant family that is storing grain for the winter. The ants refuse the grasshopper and rebuke him for wasting his summer instead of preparing for the winter. The fable's moral is that there is a time for work and a time for play. Though the fable is over 2,500 years old, it captures the most common and consequential mistake in creating wealth: failing to save early in life and then discovering later that the opportunity has passed and the consequences are dire.

A young man explained to me that he changed his career plan after witnessing the grasshopper effect while working temporarily as a bank teller after college. He frequently saw elderly people withdraw funds from their dwindling accounts with desperation, fear, and tears in their eyes. He decided to set aside his urban planning degree and become a financial advisor to help people avoid that situation.

If the level of household debt, frequency of bankruptcies, and inadequate planning for contingencies and retirement is anything to go by, more people choose to be grasshoppers than ants. Tragedies are avoidable if you strike a healthy, well-considered balance between immediate gratification and preparing for the distant-but-certain future. This decision will become increasingly important in the New Reality.

Acquiring wealth instead of managing it

Virtually everyone focuses on income and wealth acquisition, but few pay attention to managing wealth. Many people assume that their income or assets (or lack thereof) do not justify wealth planning. Other common reasons for inattention include taking current financial security for granted and becoming lax or intimidated by financial matters as we age.

Even well-educated, highly paid, successful professionals and executives can lack basic personal finance knowledge that causes them to make naïve or unwise decisions. I have witnessed the tragic results dozens of times, as illustrated by the following three examples:

1. Laura was a charming, hot-shot CPA who became a corporate turnaround executive. She retired at age fifty-five with a net worth in the top 1 percent of people in the United States. Ten years later, her net worth had evaporated due to the carrying costs associated with her excessive personal real estate investments. Ultimately, she was forced to liquidate her real estate portfolio in a down market to avoid bankruptcy.

2. Ned had an Ivy League MBA and a twenty-year executive career with earnings exceeding $500,000 annually for the ten years prior to his company being unexpectedly sold. Suddenly, Ned lost his difficult-to-replace job. The fixed costs of his luxury lifestyle largely funded by debt led to personal bankruptcy within twelve months.

3. Armand's CEO career created the net worth to support a dream retirement for him and his wife, June. Five years later, however, their retirement dreams were shattered by losses in speculative, high-risk investments. Armand had incorrectly assumed that his knowledge and skills as a CEO qualified him to be a venture capital investor.

∽

On the other hand, individuals with so-called "modest" incomes and careers can build wealth securely for themselves and their families if they exercise forethought.

Claire is a single mom with three children who worked as a customer service representative her entire career. She never earned more than $50,000, and less for much of her career. But Claire's father taught her while in high school to invest 10 percent of every paycheck into mutual funds. Claire sometimes worked overtime or took a second job to support her family or to add to her investment accounts, but she never withdrew money from her investments whether the stock market went up or down or if she really needed the cash. Today, Claire's children are all college graduates with good careers and strong families. She owns her simple but comfortable home outright, has a new car, and was able to retire at age sixty-five with more annual income than when she worked. Claire's story is not uncommon, as a simple Internet search will demonstrate.

Wealth-Creating Fundamentals

Everything in the previous chapter about optimum health applies to optimum wealth.

> The consequences are dire for inadequate attention to wealth in the New Reality.

Unfortunately, a goal that is too distant—like retirement or saving for future contingencies—is often insufficient motivation to start building wealth earlier in life. It is better to identify compelling reasons to accumulate wealth, save for it, and pay attention to well-established documented fundamentals and a licensed, qualified investment advisor.

The fundamentals outlined in this section are easily remembered and apply to almost every personal financial situation, regardless of the amount of current earnings or wealth. There is an abundance of literature available for anyone who would like to do their own research. Your financial advisor can provide their own summary or interpretations of strong fundamentals. Those explained here are intended as a starting point.

Time value of money and compounded interest

The longer that money can be invested, the larger the stock market rise over time and the greater the possible investment risk/return. Few non-financial people truly understand the power of compounded interest and the time value of money to build wealth. Your financial advisor will be able to provide many examples, but for now, let's consider just a few simple ones.

First, Mary begins saving $1,000 per year at age thirty, while Sam starts at age forty. By age sixty-five, Mary will have twice as much saved as Sam. Second, investment returns over time can equal or exceed the amount invested. For example, $200 per month invested over twenty-five years would result in a total of approximately $120,000. Almost half of the amount would be the contributions that were made. The other half were produced by time and compounding of interest. Third, even a one-time investment works magic. A single $5,000 investment would grow to $28,160, assuming a 5 percent compounded annual growth rate.

Dollar cost averaging

Dollar cost averaging* works in tandem with the time value of money. Equal amounts are invested at fixed intervals, such as

---

* According to Investopedia, dollar cost averaging is an investment strategy in which an investor divides up the total amount to be invested across periodic purchases of a target asset in an effort to reduce the impact of volatility on the overall purchase. The purchases occur regardless of the asset's price and at regular intervals. In effect, this strategy removes much of the detailed work of attempting to time the market in order to make purchases of equities at the best prices. Dollar cost averaging is also known as the *constant dollar plan*.

monthly or weekly, relying upon the total stock market's inevitable rise instead of trying to guess the best time to buy or sell. Dollar cost averaging is an achievable life-changing discipline for everyone, but especially for small investors and young people.

Tax-favored investments

Investments that allow taxes to be deferred or prepaid can dramatically increase compounded interest and time value of money. They create wealth more effectively than any other investment for many people. Yet a surprising number do not take advantage of the amazing wealth-building opportunity of 401(k) accounts, investment retirement accounts (IRAs), and Roth IRAs (in the United States). Employer matches to employee contributions to a 401(k) account even further increase returns and should be maximized if offered by an employer, but many employees fail to do so. Roth IRAs allow investors to pay their taxes on investment contributions up front. The investment then grows tax-free, and no tax is due when funds are withdrawn at retirement. Roth IRAs are particularly effective for investors with lower incomes and longer investment horizons such as younger people.

Diversification or asset allocation

"Don't put all your eggs in one basket" is a way of expressing diversification, which today has become almost a science. Asset allocation is the first layer of diversification. Financial advisors use asset allocation models to evaluate risk-return. "Monte Carlo" models predict the likelihood of investment returns over time. Both models are extremely useful in setting and adjusting financial goals and plans. A typical modest risk asset allocation might be 40+ percent bonds, 40+ percent stock, 10-20 percent real estate (including the family home), and 2-5 percent higher-risk investments. Your financial advisor will help identify the right mix for you. Further diversification is explained in the following sections.

Stock market investing

"Random Walk" is perhaps the most important principle of stock market investing, but many people have never heard of it. Decades of evidence prove that consistently predicting individual stock prices is impossible because stocks follow a random and unpredictable path. However, the entire stock market can be more accurately predicted to grow long-term because it reflects economic expansion.

Based on that evidence, dollar-cost-averaged investments are made in diversified mutual funds, exchange-traded funds, and other diversified holdings that usually increase over time as the entire stock market increases in value. The investor tolerates stock market ups and downs because losses are temporary, unrealized paper losses that have historically recovered in time.

Attempting to time investments to market ups and downs has been proven to be no more successful than flipping a coin, no matter how good an investment advisor claims to be. Worse, unsophisticated investors become emotionally connected to stock market gyrations. They lose money panic selling when the market declines, and lose money panic buying in a rising market.

Wall Street legend John Bogle founded Vanguard Investments on Random Walk principles and launched the entire mutual fund industry, which made investing feasible for more small investors.

Bond investing

Bonds are important to diversification because they offer low principal risk and guaranteed income. A bond is a loan to borrowers such as governmental entities and corporations. Interest is paid to the investor at specified rates and intervals until the bond matures and the issuer returns principal. Independent rating agencies rate bond issuers and specific issues to assist buyers with risk/return decisions. For example, highly rated AAA-BBB corporate bonds are reported to historically default only 0-1 percent of the time and municipal bonds 0.8 percent of the time.

Lower risk and stable income cause bond returns to be lower than stocks. A twenty-year AAA-BBB municipal bond historically paid around 4 percent tax-free annually, though returns varied with interest rate fluctuations. Because of their stability and security, bonds are often favored for retirement income. Investors are largely immune to bond price changes because the current bond price only matters when a bond is sold before maturity.

Bond purchases are usually dollar cost averaged and diversified in a "bond ladder." That is, bond purchases are made across a variety of bond issuers with maturity dates staggered over years and quarters, so that no single bond is large enough to impact net worth, if the bond issuer defaults. Staggering maturities through a bond ladder also provides a steady supply of cash for liquidity as bonds mature. A bond ladder, however, may be impractical for smaller investors. Alternatives include bond mutual funds and exchange-traded bond funds. A financial advisor can help you select the best alternative for your situation.

Financial advisors are paid more to manage stocks than bonds, so fewer advisors are good bond managers. Nevertheless, finding a good manager is essential since bonds could become important investments over your lifetime.

Real estate and home ownership

Real estate investment is also an effective way to diversify. However, the requirements of being a direct owner, investor, and manager of real estate are beyond most people's expertise, liquidity limitations, and risk tolerance. Real estate investment trusts (or REITs) and other real estate investment techniques avoid many of these issues. A financial advisor can explain real estate investment options and tradeoffs.

Homeownership has meant security and success to most people since World War II. However, it should be considered more carefully in the New Reality. A home can be a significant risk if it represents an excessive share of total assets and if liquid assets are inadequate to meet contingencies. Ideally, the family home should be less than 20 percent of net worth. The 2008 financial crisis devastated the finances of hundreds of millions of people and almost bankrupted the US financial system. It was caused by too many people owning debt-laden, high-maintenance homes that they could not afford.

An owned home is a fixed expense instead of a true investment. However, growth in home market value while reducing a mortgage is the most common method of wealth creation in developed countries. New options for home ownership will arise in coming decades, as discussed in the next chapter.

Common pitfalls

People commonly make financial mistakes such as trying to beat the odds, spending far beyond their means, failing to diversify, ignoring risks to chase high investment returns, maintaining inadequate liquidity, making impulsive decisions, and paying for professional financial advice that they routinely ignore. Financial advisors have told me that these mistakes are so shockingly common that they seem to imply some type of breakdown in commonsense. It is important to be on guard to avoid these mistakes and instead focus on calm, rational stewardship to build wealth over time, as described below.

Optimal Wealth

More people aged fifty and under can attain more wealth in the next three decades than ever before in history. They do not need to earn more than they would have normally, depend on government subsidies, or win the lottery.

They need only adopt the wealth creation mindset, pay attention to the fundamentals previously described, and commit to an optimal wealth plan. Doing this can be a game-changer for individuals, families, and even society at large if widely adopted.

Optimal wealth is reached by establishing an achievable plan to fund a baseline lifestyle for life that frees you to achieve your potential.

Almost everyone has a formal or informal retirement plan. The usual goal of which is the continuation of their pre-retirement lifestyle after retirement for the rest of their life. They plan to fund this goal through after-tax income from investments, income from voluntary retirement plans, and Social Security in the United States. The amount of investment and voluntary retirement plan assets that are needed to produce the required after-tax income, in addition to Social Security, is routinely underestimated. The average person's 401(k) balance in 2019 was $106,478, according to Vanguard's 2020 analysis of over five million plans. The median balance was

$22,775.* Both these amounts are woefully inadequate to fund retirement for today, much less in the New Reality, yet many people continue to dream of early retirement without a realistic plan to make it possible.

Optimal wealth strategy

Optimal wealth addresses three elements that are missing from most people's financial plans and mindsets today:

1. A baseline lifestyle that you define as "enough" for you to be happy, successful, or at least satisfied.
2. Creating a nearer-term goal called "refinement." Refinement means moving retirement forward (creating the option for early retirement). This is achieved by accumulating enough assets to fund your baseline lifestyle for life when you are younger. It also means dropping the old vision of inactive retirement and replacing it with a completely different vision of purposely living your dreams to achieve your potential, and probably working longer, but without sacrificing the benefits of retirement.
3. A plan to achieve refinement using a simple financial model.

Baseline lifestyle

A curious but common practice today, particularly among the wealthy, is disposing of wealth and assets and downsizing of homes and lifestyle later in life. This occurs after devoting some of the

---

* Liz Knueven, "The Average 401(K) Balance by Age, Income Level, Gender, and Industry," *Business Insider*, March 1, 2021, https://www. businessinsider.com/personal-finance/average-401k-balance.

most vibrant years of life to acquire them.* The big question is:

> Why don't people stop pursuing more wealth and assets when they have enough, and redirect their energies and time to better use?

The most likely reason is that they never defined "enough." Therefore, our exploration of a baseline lifestyle begins by defining "enough" as specifically as possible.

Without a precise baseline lifestyle, those without wealth are likely to dream about having more than they need, believe it is impossible to achieve, and give up. Those with wealth, who have momentum, will continue to acquire wealth and assets long after they no longer want or need either. It is not uncommon for wealthy people to say that they accumulate more than they need in order to "give back." Author, entrepreneur, and philanthropist Ricardo Semler says, "Maybe that means we collected too much in the first place."

Your baseline lifestyle description should be stated as explicitly as possible and include disposable income, home(s), travel budget, automobiles, clothes, entertainment, etc. Classify each into categories of *must have*, *nice to have*, and *excess*, then rank them in priority. Doing so will help you distinguish between needs and wants. It can also be helpful to look back at life and compare when you were happiest or most satisfied with your lifestyle to determine if you had "enough" in a past time but failed to recognize it.

---

*   Haisten Willis, "Downsizing the American Dream: The New Trend Toward 'Missing Middle Housing,'" *Washington Post*, February 14, 2019, https://www.washingtonpost.com/realestate/downsizing-the-american-dream-the-new-trend-toward-missing-middle-housing/2019/02/13/0f6d0568-232b-11e9-81fd-b7b05d5bed90_story.html.

Refinement

Defining a baseline lifestyle and beginning *today* to fund it is a monumental step toward financial freedom and life success. Your goal should be to reach a point in the future, as soon as possible, where you no longer need to earn an income to support your baseline lifestyle. At that moment you become free in a way that only a tiny percentage of the world's population ever experience. You can continue to work if you choose, or you can now live exactly as you choose for the rest of your life without working. You control the two major variables to refinement: the cost of the baseline lifestyle you need and the priority that you assign to saving the money to support it. That formula is truly your "get out of jail" card, as in the Monopoly game.

Refinement includes achieving the baseline lifestyle, but it goes further to specifically define what you intend to do afterward. It is different from retirement, or even early retirement, as it is usually practiced. Most people have a retirement plan, but many lack a life plan to go along with it. Absence of a life plan unfortunately leads many people to inactivity and loss of meaning, purpose, and job-related relationships. As noted in an earlier chapter, medical research has repeatedly linked these negative factors to shorter lives and their opposite to extended longevity.

> Refinement is the opposite of idleness. It is the conscious, active pursuit of all that we can be during our most vibrant, contributory years.

Refinement asks you how you will use the freedom to be yourself, pursue your dreams, and live a lifestyle with far fewer financial constraints. Most of all, it asks you to explore your potential and what you can achieve with few if any constraints on your time and energy. Your answer may be to continue your career but in a different direction. It may be a different career, not selected by the income it provides but by the meaning you derive from it. It may be living a dream with no income involved. It may mean optimal health and relationships. The baseline lifestyle and refinement become enablers to achieve your potential by opening new opportunities for how to live and work.

It is important to define and refine "refinement" over time as you work to achieve it. The more precise and real it becomes, the more eager you will be to give it a try. There are interim motivating milestones along the way, such as watching debt diminish and net worth grow. Each brings with it an inner confidence and feeling of security.

Every person can achieve refinement by starting early enough, making it a top priority, and being disciplined with planning and spending. The people that I know personally who have achieved what I now call refinement (and there are many) describe it as liberating.

Refinement financial model

A simple financial model turns the concepts of optimal wealth and refinement into reality. Your financial advisor, accountant, or most college students who are good at math can create the financial model for you.

"Time to refinement" is the number of months and years until net worth (in liquid investments like bonds) can be accumulated to fund the baseline lifestyle. It becomes the most important financial goal and is produced from a model based on inputs such as the following:

1. Current age
2. Estimated baseline lifestyle expense
3. Current investment fund balance
4. Estimated annual after-tax income from refinement assets; the 4 percent rule discussed in fundamentals can be utilized
5. Debt other than homeownership
6. Dollar-cost-averaged monthly savings amount (the percentage of income available for debt repayment and dollar-cost-averaged investments)
7. Current lifestyle expenses, including ongoing costs (rent, mortgage, utilities, food, etc.); large, anticipated expenses (cars, home improvement, vacations); and anticipated major life events (tuition, weddings, etc.)
8. Other retirement assets include Social Security and voluntary tax-favored retirement plans such as 401(k), IRAs, annuities, etc.; factors include whether the plans are currently maximized and the withdrawal eligibility date for each
9. One-time additional investment contributions (such as bonuses, gifts, or sale of assets no longer valued, or home sale proceeds after taxes not reinvested in a new home)
10. Temporary increases to investments (derived from sources such as current lifestyle expense reductions, a second income, or an inheritance)
11. Other factors—items that may be unique to your situation, such as an anticipated promotion

The targeted time to refinement date and feasibility of achieving it can be tested by adjusting any of the factors listed above or others unique to your personal situation. Certain factors tend to dominate "time to refinement" feasibility, including age, the estimated baseline lifestyle expense, voluntary pension plan contributions, dollar cost averaging, total investment funds, and whether there will be continuing income from other sources after refinement. Your financial advisor or an accountant can help you tailor the model to your unique situation.

Managing for Optimum Wealth

Management of optimum wealth involves hiring the best possible team and keeping things simple. Metrics and reports should be simple enough for you to recall without reference. Key metrics include net worth, debt, income and expenses, cash in and out, and investment performance; all of which keep "time to refinement" on track.

A handful of metrics and a few simple reports should be as easy to create for your personal finances as it is in business. However, laws and government regulations attempting to protect investors have made it difficult by overwhelming investors with statements, legal notices, and disclosures.

Financial advisors provide reports and online monitoring of the assets that they manage, but they are prohibited by law from producing many of the needed metrics and reports you will need. Design or select metrics and reports that speak to you rather than relying upon those that advisors can offer. Hire a bookkeeper or accountant by the hour to create and assemble the metrics and reports if you feel ill-equipped to design them on your own.

Carefully select advisors

A financial team should include professional advisors for every material asset, expense, or risk. Seek professionals with excellent qualifications and track records who work with people with financial situations like yours. Check out candidates thoroughly to ensure they are competent, trustworthy, and a good fit for you and your circumstances. Disqualify any professional with whom you have a personal relationship unless you are confident that you can maintain the needed professional distance and hold them accountable.

An unpaid financial mentor should be a family member or a friend who has your best interests at heart and has managed wealth successfully or has a strong financial background. They can become your second set of eyes for decisions, plan monitoring, and advisor performance.

Your financial advisor is the licensed professional who assists in developing and executing your plan and selecting investment opportunities. Investment advisors and managers do not consistently produce better returns over time than investing using dollar cost averaging in diversified investments. Avoid advisors who claim otherwise. Do not be swayed by the size of the advisor's firm. Large firms do not necessarily produce higher returns than a small company and solo financial advisors and may be tempted to push their company's proprietary products. Do not be discouraged if you are a small investor. Many highly qualified firms specialize in small investors, including Vanguard, Fidelity, Edward D. Jones, and many others.

Hire a second independent financial advisor for a fixed annual fee to review the primary financial advisor's performance every one to two years and provide a written assessment. Relatively few financial advisors offer this performance review service, but some accounting firms provide the service. A tax accountant is essential because taxes become a large, complicated expense as wealth increases. Insurance agents and attorneys will be important team members depending upon your risk exposure.

Discuss every material decision in advance with your advisors. Their job is not to make decisions for you, but to define the consequences of your decisions and offer suggestions on issues that are outside your expertise. The more time you give to decision-making, the better off you will be in the long run. Rushed decisions create erratic and sub-optimal results and unnecessary disasters. Ask questions and keep asking them until you understand every decision. Follow the Warren Buffett investment rule and never invest in something you do not understand or with people you do not totally trust. Meet with your financial mentor and advisor ideally every quarter, and no less than twice annually. Meet other advisors on an as-needed basis, but ensure it is enough to build a relationship.

Steps to optimal wealth

1. Define materiality given your financial situation. Your financial advisor can assist you in this exercise. It is typically 3-5 percent of annual income or assets but could be higher or lower based on individual circumstances. Defining materiality is a way to focus on what matters rather than getting lost or frustrated by details that may not matter.

2. Use monthly savings from income to begin the march to optimal wealth.

3. Become debt free except for a home mortgage.

4. Build six to twenty-four months of income contingency reserve.

5. Maximize tax-favored retirement plans. Your advisor can recommend the best blend for your situation.

6. Build net worth by investing in the long-term growth in the stock market due to economic expansion.

7. Run the model when considering any material expense, or when any material amount of funds unexpectedly become available, to determine the impact on time to refinement.

8. A few years in advance of refinement, decide if you will continue to work and produce income, or if you will rely upon refinement income to fund your lifestyle. If you choose the latter, your advisor may recommend that you liquidate some stock holdings over time to purchase bonds or other income-generating assets.

9. Once the refinement goal is reached, your financial advisor may recommend a change in asset allocation or investments since you have achieved your primary goal and can now invest for future goals.

10. Generally, there are advantages to paying off the home mortgage and over time reducing the home to be under 20 percent of net worth. This is especially true since tax reductions for mortgage payments are no longer available for many people. Your advisors should be consulted, given the wide variation in individual situations and fluctuations in interest rates and real estate market conditions.

The Uncertain Future of Nations and the Current Economic System

This chapter thus far has suggested a different financial mindset, a new life goal of "refinement" to replace "retirement," and summarized some of the basics of financial planning and selecting a financial advisor. All has been offered to encourage you to place greater urgency and priority on building and managing wealth and financial security. This chapter would be incomplete without alerting you to developments that could result in remaking today's economic and nation-state infrastructure.

The meaning of money, wealth, and perhaps even the basic functions of nations are being reconceptualized by the potential of the digital Cloud economy. Innovations such as live and work from anywhere (WFA), digital currencies, virtual nations, crypto or cybernations,* immigrant investor programs, and e-citizenships are new alternatives that go to the very roots of national and international economic and political structures. Whether these changes take hold or fizzle, whether they exert a minor or a sweeping effect, and whether they take place gradually or swiftly remains to be seen. Nevertheless, it is important for you to keep an eye on developments.

These changes could have a profound effect on your career and lifestyle options because money, citizenship, and residency requirements are essential functions of nations and their economic systems. All are being challenged by an evolving "digital, virtual economy." Digital currencies are competing with national currencies and bonds for large investors. Nations rely on attracting investors at low interest rates to keep their borrowing costs low. Investors favor national governments that manage their finances so that they can repay the bonds and keep their currencies strong. Digital currencies do not play by the same rules.

A digital currency investor's gamble is that other investors will also invest, causing the value of the currency to rise. The surprising popularity of digital currencies is causing national governments

---

* According to *Merriam-Webster Dictionary*, this is the automatic control of a process or operation (as in manufacturing) by means of computers.

great concern. It is far too early to know if and to what degree digital currencies will play a meaningful role in the financial system, and how many investors will trust them. They bear watching and learning how they work. Citizenship is being redefined by immigrant investor programs that trade full or limited citizenship for a financial investment in the country or a fee. For instance, Estonia offers an e-citizenship program with limited rights for only one hundred euros. (That's correct. One hundred euros.) The Cayman Islands' immigrant investor program provides citizenship at varying levels and rights to people of independent means for a cost ranging from $20,000 to $100,000. Portugal has advertised citizenship on Facebook for $250,000. Most nations, including the United States, have similar programs. Smaller nations like the ones discussed here, and others such as Ireland, Dubai, and Israel, are likely to experiment with alternative residency and citizenship options to attract wealth and talent to their countries as their populations decline. If they do so, and wealth starts to flee larger nations, then the larger nations will need to respond in order to compete.

For example, Angus and Sarah are US citizens who live on a small yacht in the Caribbean. Their residency moves with them and the yacht. They have a second citizenship on one of the islands and are considering making it their primary residence. They have moved 20 percent of their liquid assets into digital currencies and are watching developments to decide whether to move more. Angus and Sarah have effectively decided on a mobile residence with a citizenship tied to a nation with a widely accepted passport and low taxation. As another example, Lisa, Terry, and their three children are US citizens living and working their way through Europe while shopping for a permanent residency in a country offering an optimal tax structure and lifestyle.

Cloud citizenship, independent of any physical nation, is also feasible.* A cybernation could offer a blockchain-secured digital passport for a fee. A passport essentially is an endorsement by a nation that you are who you represent yourself to be. The nation does not take any responsibility for your actions. A digital blockchain passport could theoretically provide far more robust security than the present passport system. Benefits would include the potential for a lifetime passport instead of renewals, prevent falsified or fake passports, and provide more reliable, cumulative information about travelers that could expand "trusted traveler programs" to simplify travel and border crossings. No digital passports exist today, and their feasibility will depend on nations accepting them. However, they remain a possibility for the future as the Cloud becomes the center of economic activity and world connectivity.

Cloud citizens with a passport could choose residence from nations with optimal security, taxes, and living conditions, as Terry and Lisa are already doing. Competition almost always improves products and services. If there were millions or tens of millions of people like Terry and Lisa who are highly desirable as residents, and their own populations were declining, national governments would be forced to become more fiscally responsible to compete with lower taxes. Otherwise, they could lose citizens or residents just when they need them most to finance their aging populations. An estimated 42,000 millionaires emigrated from France between

---

*   Bruno Maçães, "The Crypto State: How Bitcoin, Ethereum, and Other Technologies Could Point the Way to New Systems of Governance," *City Journal*, Autumn 2020, https://www.city-journal.org/technological-developments-new-systems-of-governance.

2000 and 2012 to escape a wealth tax that was subsequently repealed.* Current in-migrations in the United States from high-tax to low-tax states may be an indicator of things to come between nations. The economic principle of "capital knows no borders" will increasingly be a factor in future decisions about investments, choice of currency, citizenship, and residency.

None of these emerging developments are certain to survive or become trends. But for those likely to live three decades, they bear watching as you plan your career, lifestyle, and the financial plan to support them.

A Tale of Optimizing Wealth and Refinement

An older, wiser friend explained financial fundamentals to me when I was in my early thirties, long before I was a highly compensated CEO or attained wealth. He encouraged me to position myself for what I now call *refinement*. I defined my baseline lifestyle and my vision for a new life and made creating the wealth to finance them a priority. Doing so enabled me to leave a lucrative corporate career at age fifty-two. My dreams of travel, optimal health, and entrepreneurial ventures have been realized, as has my desire to be a more meaningful part of my children and grandchildren's lives. His advice so impacted my life's meaning that I have shared my experiences with others over the years, and now pass them onto you.

---

* Greg Rosalsky, "If a Wealth Tax is Such a Good Idea, Why Did Europe Kill Theirs?" *Planet Money*, NPR, February 26, 2019, https://www.npr.org/sections/money/2019/02/26/698057356/if-a-wealth-tax-is-such-a-good-idea-why-did-europe-kill-theirs.

CHAPTER 11
Success

> "Your time is limited, so don't waste it living someone else's life. Don't be trapped by dogma—which is living with the results of other people's thinking. Don't let the noise of others' opinions drown out your own inner voice. And most important, have the courage to follow your heart and intuition. They somehow already know what you truly want to become. Everything else is secondary."
> —Steve Jobs

Contemporary Definitions of Success and Why They Are Flawed

People aged fifty and under are concerned about how to become successful (or more successful) in the turbulent times ahead; they worry about preserving their present lifestyle while also being able to fund retirement. Sadly, some may have already given up on their dreams. At the other end of the scale, those people fifty-plus are often asked for advice on how to become successful but find it difficult to describe beyond generalizations. Everyone is challenged to understand success because contemporary definitions and models are abstract, ill-defined, or unactionable.

> After 5,000 years of civilization and the glorification of achievement, there is no clear, consistent, widely accepted definition of success.

In Fyodor Dostoevsky's classic novel *The Idiot*, the kind, intelligent, ethical Prince Myshkin is treated as a simpleton in a society obsessed with wealth and power. Upon learning that he is a prince, however, people shamelessly pursue his attention. Dostoevsky's story of values twisted in pursuit of social status is as relevant today as when it was published in 1869.

> We have been conditioned for centuries to accept society's definition of success rather than our own.

Societal definitions are constantly reinforced today in advertising, movies, television, news, and especially in social media. Self-defined success should prevail over other people's choices, but the need for societal approval remains powerful, particularly when endorsed by parents, spouses and life partners, friends, and family.

Parents play an important role in defining success for their children either intentionally or inadvertently. Some parents urge their children to follow in their footsteps; others encourage them to discover and follow their own dreams, talents, and passions; and still other parents sadly browbeat their children to become sports stars, physicians, or perhaps movie stars, so they can vicariously satisfy their own lost or imagined potential. Young adults frequently mimic their parents' careers and lives out of admiration for them, a desire to please, or because they believe that walking an already-trodden path will reduce their risk of failure. Other children reject their parents' path and values, sometimes to rebel.

Contemporary definitions of success are usually some combination of the following: wealth, fame, power, social status, and appearance. Each can be a noteworthy achievement, but they are flawed as definitions of success. Entertainment and news media, and social media glorify them as success factors, but only a tiny percentage of the population can hope to achieve them as portrayed. If you buy into their definition of success, you are setting yourself up for almost certain failure against that standard. Unfortunately, young people often unquestioningly adopt these false measures of success and chase them.

If we look at each contemporary success definition closely, we see that they are all transitory—no staying power, lost as quickly as gained. Money, power, and social status can easily be lost and fame and appearance fade. Trying to hold onto these things to justify one's success or standing in the world can make you paranoid and self-destructive. Ask yourself, *What kind of success would make me miserable?* The reality is, you can be successful and possess none of the contemporary measures.

Does someone who has racked up a series of significant achievements in their life have their success devalued because they are not wealthy, famous, powerful, high society, or good-looking? Of course not. In reality, the media's view of success equating to happiness is smoke and mirrors. You only need to read the memoirs, biographies and autobiographies, or listen to interviews of people celebrated by media to see that behind the façade often lies disillusionment, unhappiness, and self-destructive behavior.

Allowing other people to define your success is unwise in several respects. To begin with, they are generalizations rather than applicable to the unique mix of values, interests, talents, and passions that define you and your potential. Second, contemporary definitions of success are external measures—they are other people's interpretations of success—you may achieve them but find them unsatisfying.

But there is more at work here. Carl Jung said there are two ways of differentiating things and making decisions: *thinking* and *feeling*. By *feeling*, he meant that gut feeling you get when something is right or wrong. Or the feeling of patriotism you get at an event where your national anthem is played. It comes from within, and you can neither fake it nor make it happen.

> Success is an internal, wholly personal experience. It is a feeling that arises on its own and cannot be made to happen or fit anyone's definition except your own.

It is a personal, private amalgamation of thoughts, feelings, emotions, and deeply held values, like a broth that combines complex ingredients to create a distinct taste. It emerges as a state or experience of fulfillment, completeness, and satisfaction. For this reason, societal definitions of success, or those of other people such as parents, can never be wholly accurate or satisfying.

Constructing your life around contemporary definitions of success can have unfortunate consequences. Choosing a career early in life as a doctor, lawyer, military professional, or to work in the family business is a leap of faith. It may or may not make you feel successful decades later. The other route is to take life a step at a time and then opportunistically follow the most promising path. Whether people follow one path or another, it is common to question career and life choices between the age of thirty-five and fifty.

> Looking ahead to the second half of life, most people want more than what they found in the first half.

Some unfortunate people find themselves economically trapped by their current income and lifestyle and become embittered that they cannot break free. Divorces, broken families, and addictions are the tragic side effects of flawed, ill-defined, and superficial notions of success. This social phenomenon is often referred to as "middle-aged crazy" and it is very real. One of the primary reasons it happens is because people follow someone else's definition of success instead of their own.

Requirements for Success in the New Reality

Following contemporary definitions of success often leads to disappointment.

> The New Reality requires a new way of looking at success that makes contemporary definitions obsolete.

Egalitarian instead of elitist and exclusionary

Today's definitions of success are elitist and exclusionary, particularly when they focus on wealth, fame, power, social status, and appearance. The most powerful change force in the New Reality is democratization. Democratizing innovations make *everything available to everyone.* Exclusionary or elitist definitions of success will run headlong into democratizing forces that are reshaping the world far more profoundly than most people realize. Consider how elitism is failing in only two highly visible elitist cultures: Hollywood and Washington. Both were revered and glamorized in the past, but there has been a decline in interest in Hollywood award shows and an explosion of new democratized talent disrupting the Hollywood elite structure through Netflix, Amazon, and YouTube. Celebrities increasingly try to make themselves relevant by advocating worthy causes, political movements, or challenging the status quo. Political leaders were revered only a few decades ago but today seem more like barroom brawlers who insult our intelligence instead of wise, dignified leaders. Both Hollywood and Washington elites are missing the bigger democratization/disruption picture, which is that they are becoming progressively less relevant and less admired by their customers, and their behavior accelerates their declining relevancy.

The media-defined qualities of success will continue to be sought after because of their scarcity, but New Reality success will be *available to everyone.*

Defining success around a specific job, profession, or career will become impractical in the New Reality. A fulfilling career today may become unfulfilling or nonexistent in a decade or two. People will continue selecting and pursuing careers, but New Reality success cannot be based on the shifting sands of career.

> Success in the New Reality will be achieved by possessing the self-knowledge to match who you are and what you aspire to be to jobs, professions, and careers in a rapidly changing environment.

For instance, understanding how a specific career uses your natural talents, and why it stimulates your creativity or passion, is far more likely to lead to feelings of success that satisfy you and are sustainable over a lifetime. Harking back to Alisha and Susan's stories in the prologue, they each discovered what they really needed for success and then found the career and lifestyle to match. Alisha wanted a home-based career instead of a corporate position. She wanted to control her own destiny and integrate her work helping innovative companies to be successful with raising a family. Susan had a passion to travel and live all over the world and to work creatively with clients to solve problems outside of the courts. These, and other self-knowledge gems, are what you are looking to discover so that you can match them to evolving careers and lifestyles.

Current definitions of success are often associated with "the lifestyles of the rich and famous." Those lifestyles will be increasingly democratized by innovations such as rent versus own, advanced mobility, live and work-from-anywhere (WFA) technologies, personalized remote learning, remote health monitoring and treatment, or possibly even cybernations. People with wealth will continue to find ways to be elitists and pursue scarcity. However, lifestyles will not define success as much as in the past as lifestyles of the rich and famous become more available to everyone.

Decoupling success from institutions

Success in the New Reality cannot depend upon the institutions of the past. Today's frequent disruptions, extremism, and polarization will become more pervasive before they moderate. Widespread angst and pessimism can be expected as employers, communities, industries, governments, education, religion, healthcare, and social services are disrupted, and many become irrelevant or fail altogether.

> In the New Reality, resilience and a forward-looking perspective that sees change (however unsettling) through the lens of opportunity will lead to success.

Increased potential

New Reality success cannot be defined by how you see yourself today because your potential is expanding by the moment and will continue to expand through extended longevity, optimal health innovations, and integration with automation. Your personal definition of success must anticipate and incorporate your increased potential and the resistance and adversity that will accompany it.

This increased potential comes with responsibilities. Smartphones, for example, have revolutionized the way we communicate but along the way they have detracted from our humanness. How many times, for instance, do you text someone rather than call them? Visit any restaurant and you will see two people looking at their phones rather than talking to each other. Accelerating innovation will provide more and more opportunities to either enhance or detract from your humanness—the decision will be yours. The consequences of your decisions will also increase. Remember from the prologue Susan's out-of-control relationship with her AI electronic assistant Amy? Her husband, Jim, noticed that Susan had an unhealthy relationship with AI Amy who was taking charge of their lives. If you are not careful, automation has a way of sneaking up on you and making you dependent before you even realize it.

> New Reality success will depend upon maximizing your expanded potential using automation without detracting from your humanity.

More time better used

As change and innovation accelerate, so does the pace of life. Time seems less available and more precious. Competing demands for time can lead to anxiety and anxiety-driven behavior ranging from checkout line impatience to road rage. Behavior such as this will increase in frequency and severity unless we change our mindset about time and its use. *Merriam-Webster* defines *time* as "a nonspatial continuum measured in terms of events which succeed one another from past through present to future." Think of time as a desert that you need to cross, traveling from oasis to oasis. The oases are life-giving, not the desert. Time wasted is potential lost. When you use time unwisely, you feel your life and its potential slipping away.

> Lack of time is almost always a problem of prioritization. Devoting life to what matters optimizes time and reduces anxiety.

New Reality success and peace of mind will come to those who optimize their time. You can find out more about optimizing time at www.potentialistfuture.com.

Understanding resistance and adversity

As explained in Chapter 1, resistance and adversity increase proportionately to the pace of change and innovation. Adversity is a natural phenomenon that strengthens individuals and the human species by challenging survival and growth. Adversity can be one of nature's most beneficial gifts, or it can become a progressively heavier burden—depending upon how you react. Adversity can inspire or discourage, teach or baffle, helpfully slow us down or be an unremitting headwind. It can bring out the best in us or the worst. New Reality success will depend on accepting adversity's role as a pervasive, purposeful, rhythmic part of life that stimulates and guides your growth.

Wisdom in the New Reality

Wisdom is the application of consciousness to daily life. As discussed throughout this book, exponential growth in wisdom will be required to succeed in the New Reality. You will need to stay calm and forward-looking during instability, adapt rapidly, and make some of the most profound decisions human beings have ever made, unassisted by the institutions that were traditionally sources of wisdom.

This book has raised several questions and challenges that you will likely face between now and 2050. The decisions you make in response to these questions will determine how well you adapt to the New Reality and ultimately your life's quality and meaning. These questions go to the heart of what it means to be human, as the following examples illustrate. How would you go about deciding whether to extend your life or alter your appearance or body chemistry?

What factors would you weigh? Similarly, how far would you go to augment your power through medicine and automation? What will relationships mean to you when most communication and life is in the Cloud? How and where will you live as options expand and the definitions and relevance of community, states or provinces, nations, education, and religion change? How will you make decisions when everything you read and see may be falsified to manipulate you? The questions are themselves impressive, but the central question for you today is how to prepare and gain the wisdom needed to make the best possible decisions when the time comes.

Wisdom is arguably the greatest human quality because it incorporates the best of what it means to be human, including strength, perspective, compassion, decisiveness, equanimity, and unconditional love. Wisdom is not entirely age-dependent; the capacity for wisdom can be accelerated through growth, study, and practice at any age. New Reality success will depend upon developing the capacity for wisdom and applying it to everyday life.

Redefining Success for the Twenty-First Century

The best approach is to redefine success by replacing flawed contemporary definitions with your own adjusted for the New Reality. The redefinition is entirely practical and better done sooner than later—like now.

If success is defined in a way that only a few "succeed" and most fail, it guarantees disappointment and a loss of meaning. Recent books and media reports have explored the cause of widespread pessimism and dissatisfaction in an era of unrivaled abundance and promise. The causes of increased suicide rates in younger people are also being examined. There are no clear answers, but perhaps people sense a growing gap between how they currently live and contemporary definitions of success which seem out of their reach. In that gap, instead of despair, may lie the only accurate and universally achievable definition of success—a new improved us, the best that we can be. No government action, social consensus, or personal declaration is necessary.

> Each person has the choice to privately redefine success as reaching their potential. Doing so sets their life and the world on a different course.

Ancient practices.

Choosing individual potential as a path to meaningful life is not new. It is a theme that runs through ancient theologies and philosophies, including the Greeks and the European Age of Enlightenment. Throughout history, exceptional people pursued their unique potential and talents instead of what society or others dictated for them. The exceptional people that I have known personally, and referenced throughout this book, pursued their dreams and accomplished much in the process. They were extraordinarily wise and respected for how they lived. This chapter's content distills centuries-old principles and practices to introduce (in contemporary language) the discovery and achievement of potential as a definition of success. Book II of The Potentialist series is devoted to the practices of achieving one's potential and accelerating the capacity for wisdom.

Potential is a superior definition of success in the New Reality.

If we look at the most common dictionary definition of success, we can paraphrase it as "accomplishment of an aim or purpose." Potential is defined as "having or showing the capacity to become or develop into something in the future." Combining the two creates a new definition of success that reads, "the accomplishment of becoming or developing into our potential." This redefinition resolves the shortcomings of contemporary definitions, meets New Reality requirements, and democratizes success.

Success that is personalized to each person's potential provides clarity and direction in ways that general definitions such as career or the media definitions of success can never provide. Success no longer focuses on what we do (careers, jobs, professions), what we possess (wealth, fame, power, social status, appearance), or how we live (lifestyle). Success instead becomes who we are at our core and what we do with what we have been given. That is the truest, purest, most personalized definition of success.

> As the world democratizes, so must the definition of success to make it *available to everyone.* The opportunity to achieve individual potential is not limited by age or individual circumstances.

It is equally available to a fifty-year-old plumber, a twenty-year-old agricultural worker, a thirty-five-year-old Wall Street investment banker, and a sixty-five-year-old retired steelworker.

> Equating success as one's potential creates the possibility that everyone can become what we today consider *exceptional* in their own way. Even a small increase in the percentage of the global population approaching or achieving their potential raises the bar for humanity.

This is especially true as human potential organically expands through longer lives and automation.

> Living to potential could become a unifying ethos because everyone benefits from everyone else's achievement.

Potential is the distance run, the obstacles overcome, and who you helped along the way. Life becomes enjoyable, and at times intoxicating, because you are continually growing and contributing. What you achieve is important, but it doesn't need to be grand. The yardstick you should use is whether you did your best to become your best. You become the only audience that matters; your progress is confirmed in quiet moments in a taste of "success broth" as described earlier. Your flourishing potential becomes your brand, a unique light in the world.

> Equating success as one's potential relies on individual effort alone from the instant you make the decision to pursue your potential relentlessly, until your last conscious moment. You discover who you are, imagine what you can be and achieve, define your goals clearly, and persist in making them reality.

Living to your potential is natural

Nature is always changing, evolving, diversifying, and experimenting in a never-ending drive to perfect the individual and the species. You are part of that system and can learn from it. Every living thing is engineered to strive for its potential and serve the species in the process. Nature creates a yearning so strong that a tree can grow through rock. Disabled athletes double-down with superhuman drive to compete.

That relentless motivating aspiration resides in each of us as it does in all living things. But we alone have free will; we must choose to live to our potential and contribute to our species. You will feel different once you make the decision to reach your potential—life will feel different. It is as if a natural underlying process begins to encourage you. You have probably experienced this in the past, when everything in your life was headed in the right direction and positive developments came out of nowhere. This is often described as "being on a roll," "in the zone," or "hitting on all cylinders." Those feelings emanate from the alignment of that natural drive with the path to your potential. Seeking and repeating that experience is an invaluable navigational aide throughout life. So is its opposite. When nothing is going right for you, a course change is required.

Equating success with potential creates the possibility that everyone can become exceptional in their own way.

It is likely that only a small portion of the population reached their potential in the past. My experience meeting and getting to know people, in many parts of the world, is that most are well-meaning and want the best for themselves, their families, and communities. But "exceptional" is a word that rarely applies. Meeting extraordinarily wise, productive, confident, directed, compassionate people is uncommon. The difference between being well-intentioned and being exceptional, in my experience, is choosing potential as success. Making the committment to discover and achieve your potential is far easier and more practical than you might imagine.

The Potentialist

Pursuing potential as success begins by visualizing your goal as clearly as possible and keeping it in the forefront of your daily life, then activating processes and practices that aggressively help you pursue it. There is no single or prescribed way of doing this; however, the method summarized here is the one I have used. It is derived from literature on the subject and the practices of exceptional people I have known.

"Potentialist" is the name I assigned to the pursuit of potential as a central theme in life. The word became my mental point of orientation, like how other people may define themselves by their career, hobbies, or something else. The word "Potentialist" is packed with meaning. I created a twenty-seven-word sentence as a reminder.

"I will do my best, to be my best, and leave the world and those I meet along the way a little better than I found them."

Using this sentence as a daily commitment keeps potential in the forefront of my mind and guides me through each day. The words capture what nature asks of all living creatures. They reflect recurrent literary and philosophical themes of potential as a path to wisdom and meaning in life. The sentence is memorable and instructive to me—I hope it will be to you.

If we look at the component phrases, we can clarify the meaning and call to action.

"I will do my best . . . "

There is an emotional and physical certainty in the words "I will do my best," or "I have done my best." You alone recognize when you have given your all and there is nothing left to give. Your best is all that life can ask of you and all that you can expect of yourself. Maybe nature provides this guidance, so we know when to continue and when to stop without remorse or second-guessing. People who believe that "doing your best" is unachievable or unknowable are likely driven by perfection or haunted by the expectations of others. Exceptional performance is doing your best; it's not perfection.

Pushing to find and then respect your limits is a healthy way to do your best. This is something every athlete, physician, hard-driving entrepreneur, executive, or other professional understands:

"To be my best."

You discover your best self through a lifelong process of unfolding, expanding insights and perceptions. You self-discover, but equally you learn from other people's insights in conversations, shared experiences, and literature. Some of the richest learning opportunities are the candid insights from those who know you well and want the best for you.

You can discover your deepest, truest, most complete self through your interests, passions, talents, faults, and excesses. The more objective and non-judgmental you are with yourself, the more you will grow. Objective assessment defines your baseline and aspirational self, in a process like personal brand development and improving relationship skills in earlier chapters.

Changes to your perspectives, mindset, and behavior can be identified and tracked from a predetermined baseline to your aspirational goals. These changes are best prioritized so that you achieve small victories, that in time grow to larger ones. The process must be given the formality and time it deserves. It needs to be written, at least in summary, with scheduled priorities and regular review.

As you grow, your choices become conscious and deliberate instead of instinctive, unconscious, or achieved by trial and error. In time, you will discover the 99 percent rule: choices to pursue growth and potential will be helpful 99 percent of the time, while other choices will stunt growth or result in a negative outcome 99 percent of the time.

Self-discovery and pursuit of potential is often misunderstood. Some people believe that it is narcissistic. Others think it is idle navel-gazing or idealistic aimlessness. In fact, it is the opposite. Committed pursuit of potential is a deliberate, aggressive, directed path through life.

Exceptional people throughout history led fulfilled lives and made contributions that improved the lives of millions because they walked their own path and relentlessly sought the best from themselves. Millions of exceptional people have made untold contributions but are by necessity overlooked in history books. Similarly, the exceptional people that I have known were highly committed contributors who lived by the principle that time wasted is potential lost.

> Perpetual lifetime growth in search of your best self becomes a way of living.

Time commitment is insignificant and unnoticeable because it becomes an integrated expansion of how you think, observe, and act. It is not new effort, but current effort better spent. It is not always easy, but the process is exhilarating.

"Leaving the world and those I meet along the way a little better than I found them."

Pursuit of your individual potential alone will fail to produce the inner satisfaction with life that we all seek from success. You must also contribute to others, and life in general, in some way meaningful to you personally. Satisfaction will elude you if your contribution stems from a desire to be admired or to feel generous or righteous. The commitment expressed in this phrase is far more meaningful than displaying periodic generosity or kindness. You must commit to living your life to its fullest while intentionally, *in every incidence large and small,* leaving the world and every person you touch better than you found them. Reading that back, I assure you it sounds more daunting than it is in practice.

Everyone's actions are limited to the knowledge and capacity for wisdom that they have at any given time in any specific situation. Understanding and accepting that reality is a critical step in maturity, equal to accepting that you are limited to doing your best. Acceptance frees you from the destructive cycle of perfectionism and self-deprecation that will retard your growth and create unrealistic expectations of yourself and others.

> It is not possible to immediately leave every person and situation better than you found them. But if you act with good intention and with as much love as you can muster in the moment, you will have done your best. You will get better with practice.

Your task should be to make increasing numbers of encounters positive over time, and to break patterns where you are the cause of negative outcomes.

Living as a Potentialist could seem like an endless commitment without receiving some sort of reward. There is no perfect individual or species in nature; we all have room for individual improvement. Fortunately, living for your potential changes how you experience life and reinforces the pursuit with rewards along the way.

As I mentioned earlier, success is an internal state of fulfilment or satisfaction that arises from within—it is something we do not control. Long before you are fully living your potential and before success feels complete, two incredibly satisfying interdependent milestones emerge that signal that your potential is just ahead. Those achievements on the road to potential are the dignity of self-achievement and expanded perspective through self-discovery.

Think of them like new muscles that provide previously unknown strength and new energy that you didn't know you had. Like intertwined strands of DNA, they unify previous divisions or contentions in life such as public and private self, career and life, and perceptions and decisions based on "both" instead of "either/or."

The dignity of self-achievement derives from honorable success in career and life. Every human being craves this experience. Without it, life is perpetually unfulfilling. With it, people become free and proud. You can observe the drive for the dignity of self-achievement in very young children. Nothing makes a three-year-old child happier than to achieve something on their own. You will often see them angrily rejecting help, usually saying something like, "Leave me alone; I want to do it myself!" This incredibly important developmental step does not end with childhood but continues throughout our lives. Parents, well-meaning people, and government programs sometimes fail to appreciate that satisfying this need is essential to human development and dignity. Robbing a person of it, even with the best intention, condemns them to dependency and loss of meaning in life. Lynne Twist describes the tragic effects in her book, *The Soul of Money*: "Much as the organized beggars of Bombay had learned to present themselves advantageously for alms, this charity relationship based on pity and sympathy for 'the needy' began to show up for me as a kind of pornography of poverty that demeaned all parties. I have seen its cost again and again in my work in the developing world. I see people with a dependency hangover. I see the consequences of a welfare state worldwide that goes beyond rich and poor, that is actually inside of institutions, families, nation-to-nation relationships where people 'help' other people in a way that is patriarchal—from the top down—and creates dependents, and dependence, instead of supporting self-reliance and healthy interdependence. It diminishes everyone." Contributing to others without harm requires the maturity and wisdom to know where that line is, as every parent knows all too well.

Expanded perspective through self-discovery emerges as the intertwined partner to the dignity of self-achievement. Expanded perspective sharpens our perception of the world around us and our reactions to that world. Both are necessary to expand wisdom. Expanded awareness fuels a desire to understand life and how things work as never before. That desire in turn leads to heightened curiosity, awe, respect, and gratitude for life.

Life for me became a beautiful mosaic of gemstones of varying sizes. Everything I learn and every person I meet is a new gem piece. Every discovery intensifies the desire to uncover another. Each gem piece, put in its correct place, expands, deepens, and sharpens the perception of life's magnificent mosaic. Increased perception clarifies your place among, and connection to, all else in life in a way previously unimaginable.

The dignity of self-achievement and expanded perspective through self-discovery serve the dual purpose of being milestones and prerequisites to achieving potential. Their appearance signals an expanded capacity for wisdom that when applied to life becomes our greatest gift to others.

The ultimate reward for success through our potential can be expressed in another simple but powerful statement:

"There will be more to do until my last breath, but if this is my last, I am complete."

In these words we commit to growth and contribution until our death.

The word "complete" explains what the entire race of life has been about. Completeness is the ultimate reward for a life well-lived. Death perceived as premature extinction is replaced by completeness as promise fulfilled.

EPILOGUE

THE PROLOGUE INTRODUCED YOU TO Alisha and Alejandro, Greg and Linda, Peter and Ed, and Susan and Jim, among others. Many of the book's characters are based on real people with names and situations altered to protect their privacy. Others are composites of real people, and a few are totally fictional where it was not possible to project current people's circumstances into the future. Whether real or imagined, all the characters in this book, and all the exceptional people referenced throughout, strive for their potential while leaving the people they meet and the world better than they found it. It takes both dreams and actions to live such a life.

Those living through the next three decades have an opportunity to experience the best, most meaningful lives in human history. If you are one of them, all you need to do is seize the opportunity. You can *live longer and healthier* by preventing many of the current illnesses and disabilities that currently plague us, and by utilizing optimized health innovations that will becoming increasingly available and affordable. You can *live better* through expanded lifestyle opportunities, more meaningful work, and a reduction in tension between work and lifestyle. You can be *more knowledgeable* by taking advantage of personalized distance learning and the intelligent use of technology and unlimited information. You can be *more powerful* as you become augmented by automation and live a longer, productive life. Your life can be *freer and more meaningful* by being enhanced by optimal wealth and by pursuing your potential. You can be *wiser* by developing the capacity for wisdom earlier in life, by living longer and enjoying more varied experiences, by making more and better choices, and by assuming greater individual responsibility.

The most successful people in the twenty-first century will have come the farthest from where they began. They will live to their potential, using all their talents and energy to overcome barriers and negative behaviors, and to become the best they can be. They will achieve and create independently but collaboratively and leave the world and the people they meet better than they found them.

INDEX

B

multiple personas, 124

Musk, Elon, 15–16, 98n10, 100, 155, 160, 165

Myers-Briggs Type Indicator (MBTI), 121

N

NASA, 99

nature, 226–227

Netflix, 48, 50, 56, 217

Neuralink, 98n10

New Reality, 28, 32–33, 36
availability of personal information in, 114–115
careers, 92
"changes to work" concept, 78–79
"Cloud-dependent" concept, 81–82
financial mindset, 188
lifestyle integration, 80–81
modern-day communications technology, 76–77
offering opportunity for health, wealth, and success, 169–170
positives and negatives, 84–88
re-evaluation of priorities, 79–80
relationship-building, 77–78
requirements of success in, 217–223
"superhuman" concept, 75–76
sweeping societal changes, 83–84
travel, 68
"trial and error and reacting" concept, 82–83
"who we are and how we live" concept, 74–75
wisdom in, 222–223

New Reality relationships, 133–134
addressing individual differences, 141–142
developing healthy relationships with automation, 139–141
inadequacy of relationship skills, 135–136
kinship goal, 143–145
kinship prerequisites, 145–149
receptivity to new forms of communication, 138–139
shared experiences and empathy, 142–143
twenty-first-century relationships, 136–138

news media, 49, 52, 54–55, 87, 215

New Work, 79, 91, 92
features, 92
pacing factors for change, 93–94
workforce transition, 94–96

Nin, Anaïs, 113

non-military drones, 70

novice entrepreneur, 163–164

O

objective assessment, 229

Oculus VR, 70

on-demand ride services, 38

one-size-fits-all instructional method, 54

online cash transfers, 57

open-source technology, 156

optimal health, 174–175. *See also* health mindset
case study, 185–186
customization, 184
draft baseline, 184–185

collaborative personality attributes development, 103–105

collaborative products and services, 99–100

finding leading-edge adaptive and collaborative organizations, 105–109

insights from thought leaders, 98

mass production of problem-solving and innovation, 96–98

people as problem-solver and innovator, 99

rethinking career game plan, 109–111

TED Talks, 50

Teresa, Mother, 152

Tesla, 105–106, 164

Texas A&M-Commerce, 3

TFR. *See* Total Fertility Rate (TFR)

thinking, 158, 216

critical, 86

like entrepreneur. *See* entrepreneur(ship)

polarized, 60

thought activation, 21

TikTok, 50

time

investment in, 121

management, 183

role in success, 221–222

value of money, 194, 195

"time to refinement" feasibility, 203–204

Toffler, Alvin, 13, 32, 72

Tolstoy, Leo, 2, 4

Total Fertility Rate (TFR), 64–66

total health. *See* holistic health

"trial and error and reacting" concept, 82–83

Tripadvisor, 56

Twain, Mark, 151

Twist, Lynne, 189–190, 232

U

Uber, 38, 99

unhealthy lifestyles, 173–174

United States, 64, 65, 175, 212

for entrepreneurship, as place for, 155

need for health plan system in, 158–159

social security in, 199

US Medal of Honor, 101

V

V10 social media, 110n11

value

of being, 32

of currency, 209–210

proposition, 59–60, 107, 111, 152, 153, 155, 160

in relationships, 136

shared, 147–148

virtual nations, 209

virtual reality, 12–16, 27, 97, 102, 110n11, 136, 142

virtue, 3, 88

volatility, 32, 43–44, 72

VRBO, 18

W

Wall Street Journal, 156

War and Peace (Tolstoy), 2

war, democratization of, 51